InSAR and Deep Learning in Landslides Research: Intelligent Identification, Risk Assessment and Susceptibility Mapping

Yi He

InSAR and Deep Learning in Landslides Research: Intelligent Identification, Risk Assessment and Susceptibility Mapping

Springer

Yi He
Faculty of Geomatics
Lanzhou Jiaotong University
Lanzhou, China

National-Local Joint Engineering Research
Center of Technologies and Applications
for National Geographic State Monitoring
Lanzhou, China

Gansu Provincial Key Laboratory
of Science and Technology in Surveying
and Mapping
Lanzhou, China

ISBN 978-981-96-9131-9 ISBN 978-981-96-9132-6 (eBook)
https://doi.org/10.1007/978-981-96-9132-6

This Springer imprint is published by the registered company Springer Nature Singapore Pte Ltd.
The registered company address is: 152 Beach Road, #21-01/04 Gateway East, Singapore 189721, Singapore

For an intelligent recognition of landslides, and for alleviating their global toll.

Preface

Landslides refer to surface slope rock and soil masses that rely on their inherent weak structural planes. With the intensification of global climate change, landslides occur frequently and cause severe damage. Interferometric Synthetic Aperture Radar (InSAR) technology combined with artificial intelligence has become an important means for landslide monitoring and assessment. However, key challenges remain, such as unclear fusion and correlation mechanisms for multi-model and multi-source heterogeneous data in complex and hazardous environments, as well as uncertainties in landslide susceptibility prediction. Therefore, revealing the intrinsic correlation mechanisms between multimodal data and landslide characteristics, constructing intelligent coupling models that capture the spatiotemporal evolution patterns of landslides, and achieving efficient, high-precision, and interpretable intelligent landslide identification and prediction are urgent scientific problems to be addressed. These advancements are of great significance for enhancing national landslide risk monitoring and early warning capabilities, safeguarding people's lives and property, and supporting sustainable development goals.

The content of this book primarily focuses on two aspects: precise intelligent landslide identification and intelligent susceptibility mapping/assessment. The main contributions include:

(1) A multi-source feature cooperative optimization and multi-model integrated intelligent landslide identification method was proposed. By combining spectral features, topographic features, SAR image polarization features, and texture features, a collaborative mechanism for global characterization of multidimensional information "spectral response-topographic representation-texture interpretation"—was established. An intelligent landslide identification framework integrating multiple models such as U-Net, FCN, and convolutional neural networks was constructed. The contribution weights of spectral, topographic, and texture features in landslide identification were quantified, enabling adaptive focus on landslide features and suppression of redundant information, thereby achieving high-precision landslide datasets.

(2) A landslide susceptibility mapping/assessment method constrained by geographical environmental similarity was proposed. By introducing graph structures, a directed weighted graph constrained by geographical environmental similarity was constructed, capturing the complex spatiotemporal correlations between landslides and disaster-pregnant environments. A cross-space sampling strategy based on geographical environmental similarity was proposed, ensuring the balance of positive and negative landslide samples and the validity of regions with limited samples. A graph neural network model was established, effectively addressing the heterogeneity challenges in landslide susceptibility assessment in complex environments.
(3) A landslide risk assessment method coupling static and dynamic spatiotemporal features was proposed. The spatiotemporal mutual feedback mechanism between static and dynamic factors was revealed, and a multi-model integrated landslide risk assessment framework incorporating static and dynamic factors was established. Parallel artificial intelligence models were constructed to mine deep spatiotemporal features. The landslide index was optimized by integrating landslide susceptibility evaluation, InSAR based ground surface deformation, and rainfall data. Combined with the landslide vulnerability index, fine-grained classification of landslide risk levels was achieved. The risk identification accuracy of the proposed method improved by an average of over 10% compared to internationally advanced methods.

This book will provide researchers engaged in global landslide identification and susceptibility mapping/assessment with methodologies and approaches for intelligent landslide recognition and susceptibility mapping, offering data and techniques to advance global landslide research.

Lanzhou, China Yi He

Acknowledgments The creation of this book is a collective effort rather than an individual achievement. While only the author's name appears on the cover, its success owes much to the dedication of many behind the scenes. First, I am deeply grateful to my family for their unwavering understanding and support throughout this demanding journey. Second, my sincere thanks go to the research team—Wenhui Wang, Qing Zhang, Long Jin, Hesheng Chen, Binghai Gao and Zhan'ao Zhao—for their expertise in experiments, data analysis, and manuscript refinement. Lastly, this work was made possible by funding from: National Natural Science Foundation of China (Grant No. 42201459 and 42471471), Gansu Science and Technology Program (Grant No. 23JRRA881), Gansu Province Joint Research Fund Program under Grant 24JRRA855, and Lanzhou Youth Science and Technology Talent Innovation Program under Grant 2024-QN-12.

Competing Interests The author has no competing interests to declare that are relevant to the content of this manuscript.

Contents

About the Author

Dr. Yi He has been in Faculty of Geomatics, Lanzhou Jiaotong University as a Professor. He focuses on research of landslides based on remote sensing technology and deep learning. His recent research interests include deep learning landslide identification and susceptibility mapping, InSAR technology ground surface deformation monitoring and time series prediction, and key technologies of InSAR data processing based on deep learning, etc. He has presided over many national and provincial projects, including the National Natural Science Foundation of China General Project, the National Natural Science Foundation of China Youth Science Fund, the China Postdoctoral General Fund and other important projects. Since 2011, Dr. He has published 100 peer-reviewed technical papers in international journals and conferences. He is a member of IEEE and served as a reviewer of several international journals, e.g., International Journal of Applied Earth Observation and Geoinformation, IEEE Journal of Selected Topics in Applied Earth Observations and Remote Sensing, International Journal of Remote Sensing. His major awards and honors include: the 2024 IEEE JSTARS Outstanding Paper, National Railway Youth Science and Technology Innovation, Second Prize of Gansu Province Science and Technology Progress, Second Prize of China Surveying and Mapping Science and Technology Progress, et al.

Chapter 1
Introduction

1.1 Background

Landslide refers to various processes that cause slope (including rocks, soil, artificial fill, or mixtures thereof) to move downward and outward (Fig. 1.1) (Cruden, 1991; Highland & Bobrowsky, 2008). This phenomenon is a natural surface process that predates the appearance of Homo sapiens on Earth. However, millions of years later, these natural events have become natural disasters threatening human existence (Alcántara-Ayala, 2002). A quick search for the term "landslide" on Google yields 126,000,000 results. This is because the topic of landslides has attracted the interest of researchers across many scientific disciplines, including sedimentology, oceanography, geomorphology, volcanology, seismology, glaciology, Martian geology, deep-sea structural engineering, highway engineering, soil mechanics, climate change, sea-level rise, natural disasters, and petroleum exploration and production (Shanmugam & Wang, 2015). Landslides pose significant threats to human survival, development, and the environment, and research on landslide identification, mechanisms, and prediction has never ceased.

1.1.1 Risks and Effects of Landslides

Natural disasters are a series of events beyond human control that poses significant hazards to humanity, and inappropriate human activities often indirectly exacerbate their impacts. Some well-known disasters, such as droughts, storms, floods, earthquakes, volcanic eruptions, and extreme temperatures, are more prevalent. However, landslides rank as the seventh deadliest natural disaster (Herath & Wang, 2009), accounting for 17% of all fatalities caused by natural hazards (Kjekstad & Highland, 2009). From 2003 to 2010, the annual global death from landslides showed

Y. He, *InSAR and Deep Learning in Landslides Research: Intelligent Identification, Risk Assessment and Susceptibility Mapping*, https://doi.org/10.1007/978-981-96-9132-6_1

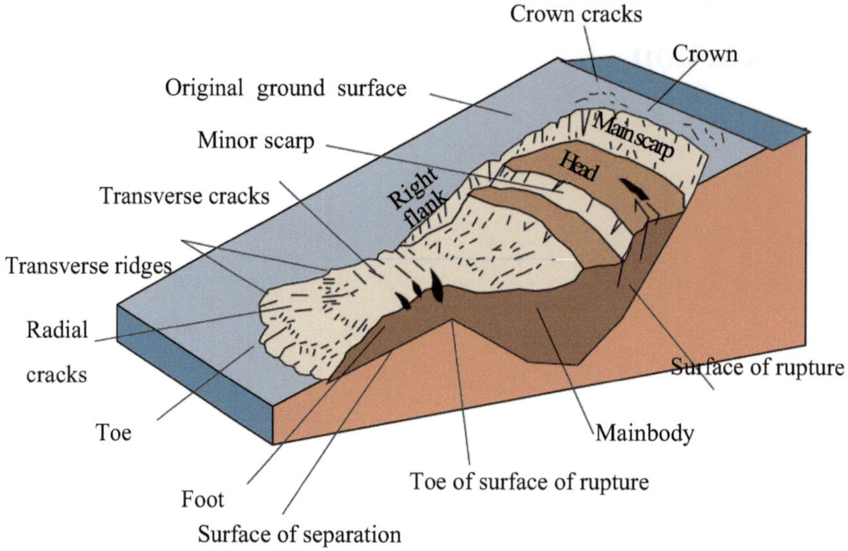

Fig. 1.1 Diagram of a typical landslide

an increasing trend (Petley, 2011). According to World Bank (WB) data, approximately 3.7 million square kilometers of land worldwide are prone to landslides, with 300 million people living in these high-risk areas (Agwe et al., 2005). This indicates that landslides threaten vast regions and populations, causing severe damage to urban infrastructure and transportation systems, as well as significant loss of life and property. For instance, the World Health Organization (WHO) reported that between 1998 and 2017, landslides affected around 4.8 million people globally, resulting in over 18,000 fatalities (Pascaline & Rowena, 2019). The economic losses caused by landslides worldwide surged from $50 billion annually in the 1980s to nearly $200 billion per year between 2004 and 2014 (World Bank, 2014).

Globally, landslides are concentrated in Asia (the region with the highest frequency, accounting for 75% of all landslides), South America, and areas such as Turkey, Iran, and the European Alps (Fig. 1.1–1.2) (Froude & Petley, 2018). China is particularly vulnerable to landslides. According to WB estimates, from 1980 to 2018, China experienced 20,950 earthquake-induced landslides (the highest globally) and 35,280 rainfall-triggered landslides (the second highest) (Redshaw & Bottomley, 2020). As reported in the *2023 China Natural Resources Bulletin* issued by China's Ministry of Natural Resources, the country recorded 3,668 geological disasters in 2023, including 925 landslides, constituting approximately one-quarter of the total.

China's complex geological environment, with mountainous terrain covering over 65% of its land area, makes it highly susceptible to landslides. Regions such as the Loess Plateau, western Sichuan, the Three Gorges Reservoir area, western Yunnan,

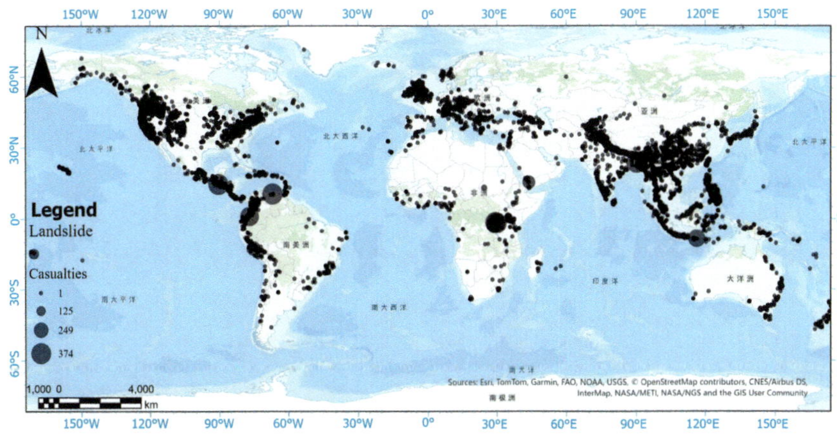

Fig. 1.2 Global landslide distribution map (NASA Cooperative Open Online Landslide Inventory, https://maps.nccs.nasa.gov/arcgis/apps/MapAndAppGallery/index.html?appid=574f26408683485799d02e857e5d9521, accessed on April 14, 2025)

and southeastern Tibet are particularly prone due to steep topography, collapsible soils, uneven rainfall distribution, seismic activity, human disturbances, and extreme precipitation events (Li et al., 2018; Lin & Wang, 2018; Xu et al., 2023) (Fig. 1.3).

Recognizing the significant threats that landslides pose to both economic stability and human safety, Chinese scholars have conducted extensive research on various aspects of landslide studies, including triggering mechanisms (Xu et al., 2023), monitoring techniques (Chen et al., 2024a, 2024b), susceptibility assessment (He et al., 2022, 2024; Huang et al., 2018), preventive measures (Choi & Cheung, 2013; Xiao et al., 2015), and post-disaster rehabilitation (Zhou et al., 2015), with particular emphasis on monitoring and susceptibility evaluation as the most critical components for effective landslide prevention and control, where early detection and proper mitigation measures serve as essential approaches to landslide management and loss reduction.

1.1.2 Limitations of Traditional Landslide Identification and Mapping Methods

Traditional landslide identification primarily employs field surveys and investigations, interpretation of remote sensing data, as well as pixel-based, object-oriented, and machine learning approaches. Field surveys represent the most conventional and direct method for landslide detection, allowing investigators to closely observe morphological characteristics and delineate various landslide elements (Li et al.,

Fig. 1.3 A landslide occurred in Tianbao Group landslide, Shiban Village, Ganlong Town, Songtao County, Guizhou Province, China, on July 8, 2020

2016; Xu et al., 2015). While this approach yields highly accurate landslide information, it requires substantial human, material, and time resources, making large-scale coverage challenging (Jiang, 2023). Additionally, field survey results require compilation and analysis, preventing real-time updates and potentially leading to information delays that hinder timely identification of new landslide (Zhao, 2024). Moreover, various secondary hazards caused by landslides pose significant risks to field investigators (Chen et al., 2024a, 2024b).

The visual interpretation method involves professionals identifying landslides based on features such as tone, texture, shape, and spatial distribution in imagery. For uncertain cases, interpreters may incorporate supplementary data like digital elevation model (DEM) and geological information. Petschko et al. (2016) utilized Light Detection and Ranging (LiDAR) and Digital Terrain Model (DTM) derivatives to conduct visual interpretation and mapping across a 19,000 km^2 study area. Following the 2008 Wenchuan earthquake, Xu et al. (2009) interpreted approximately 46,560 secondary geohazards (including collapses, landslides, and debris flows) across 14 severely affected counties using multi-source remote sensing imagery, DEM data, topographic maps, and digital geographic maps. While visual interpretation avoids complex algorithms and typically achieves high accuracy without hazardous field work, its reliance on interpreter expertise and substantial time investment results in low efficiency, making it unsuitable for emergency response needs (Li et al., 2022).

Pixel-based landslide identification analyzes individual pixel characteristics in remote sensing imagery, employing classification algorithms to determine whether pixels belong to landslide areas. Zhao et al. (2017) integrated various difference images (including band reflectance and vegetation index differences from dual-temporal Landsat-8 data) to map coseismic landslides from the 2015 Nepal earthquake using pixel-based change detection, demonstrating high identification accuracy. Yang et al. (2013) applied MODIS Normalized Difference Vegetation Index (NDVI) time-series data to detect vegetation changes and identify landslide locations in the Wenchuan earthquake zone, yielding spatial patterns consistent with high-resolution imagery and field surveys. However, pixel-based methods often produce significant false positives/negatives due to their neglect of geometric features and contextual information, limiting practical applications.

Unlike pixel-based approaches, object-oriented methods segment images into homogeneous objects (where entire landslides are treated as pixel collections rather than individual pixels), then classify them using spatial, shape, texture, spectral, and contextual features, achieving superior accuracy. Blaschke et al. (2014) developed a semi-automated object-oriented method combining NDVI, brightness, texture features (from IRS-ID and SPOT-5), with slope and flow derivatives (from DEM and GLCM), achieving 93.07% accuracy in northwestern Iran. Keyport et al. (2018) compared pixel-based and object-oriented methods using aerial imagery in Guatemala, demonstrating the latter's higher accuracy, fewer false positives, and strong consistency with manual mapping. Nevertheless, object-oriented methods face limitations in automation due to subjective parameterization (e.g., segmentation scales).

Current machine learning methods (e.g., Random Forest, RF; Support Vector Machine, SVM; Artificial Neural Network, ANN) are often integrated with pixel/object-based approaches. Chen et al. (2017) combined object-oriented analysis with RF to develop a landslide mapping framework using ZY-3 satellite imagery in the Three Gorges Reservoir area, achieving high accuracy through feature selection and mathematical morphology. Ghorbanzadeh et al. (2019) compared ANN, SVM, and RF performance using RapidEye data in Nepal. However, these methods require manual feature extraction and scenario-specific parameter tuning by experts, exhibiting poor transferability. Consequently, traditional machine learning struggles to meet demands in complex environments or high-precision applications.

1.1.3 *Application Potential of InSAR Technology and Deep Learning*

1.1.3.1 Application of InSAR Technology in Landslide Identification and Susceptibility Mapping

Microwave remote sensing, with its shorter wavelength, demonstrates strong penetration capabilities and high resolution as it is less affected by scattering or absorption from clouds, fog, or atmospheric conditions, enabling all-weather, day-and-night surface information acquisition and making it a crucial tool for landslide identification and susceptibility mapping (Zhang, 2018). Landslide detection methods based on microwave radar imagery mainly utilize two types of information: scattering characteristics and phase data. Scattering-based methods are similar to optical image analysis, relying on radar image attributes such as grayscale, brightness, texture, and shadows for visual interpretation or change detection. However, due to the side-looking imaging geometry and less intuitive data representation of radar systems, accurately identifying landslides remains challenging (Wu et al., 2022). Phase-based methods, the dominant approach in microwave remote sensing, leverage the unique advantage of Single Look Complex (SLC) SAR images, where each pixel records both amplitude and phase information of radar signals. Through Interferometric SAR (InSAR) technology, phase differences between two or more SAR images can be analyzed to detect ground surface deformation with millimeter-scale precision, establishing InSAR as the most widely adopted technique for large-scale landslide identification (Zhang et al., 2022). For example, Liao et al. (2011) used TerraSAR-X data and D-InSAR technology to monitor landslides in the Three Gorges Reservoir area, demonstrating that amplitude and differential phase could accurately identify landslide locations, deformation patterns, and temporal ranges. Rosi et al. (2017) employed PS-InSAR technology to update landslide inventories in Italy's Tuscany region and calculated landslide distribution indices and density maps, providing deeper insights into regional landslide mechanisms. Berardino et al. (2002) proposed using interferometric processing with short-baseline image pairs to eliminate multi-scattering interference, developing the SBAS-InSAR technique, which maintains good coherence even in topographically complex and densely vegetated areas, leading to its widespread application. He et al. (2023) combined SBAS-InSAR with landslide susceptibility mapping results to jointly identify landslides in Lanzhou, northwestern China, improved landslide identification accuracy. InSAR technology is playing a critical role in landslide disaster monitoring, identification, and early warning both domestically and internationally, demonstrating practical feasibility in landslide monitoring and identification.

1.1.3.2 Application of Deep Learning in Landslide Identification and Susceptibility Mapping

In recent years, deep learning technology has advanced rapidly. Compared to machine learning methods, deep learning-based models can transform unstructured remote sensing data into structured feature information, better process complex features in remote sensing images, and extract intrinsic and deep-level features, achieving superior performance to traditional machine learning approaches. This has provided new insights for landslide identification and is gradually being applied in this field (Ghorbanzadeh et al., 2022; Huang et al., 2024; Tien Bui et al., 2020). Among these methods, approaches based on Convolutional Neural Networks (CNN) and U-Net and its variants combined with optical remote sensing imagery have been widely used for landslide identification tasks (Prakash et al., 2021; Fang et al., 2022; Bhuyan et al., 2023; Wu et al., 2024).

However, landslides often occur under adverse weather conditions and within complex geographic and geological environments, such as heavy rainfall, cloud cover, or vegetation occlusion. Additionally, when training data is relatively limited or noisy, individual deep learning models may face overfitting risks, leading to low model stability, poor generalization ability, and reduced landslide identification accuracy (Chen et al., 2024a, 2024b; Wang et al, 2014).

Methods combining CNN and U-Net with optical remote sensing imagery are frequently employed for landslide detection and identification. For example, Wang et al. (2019) used CNN combined with 9-band Gaofen-1 imagery and NDVI indices to identify landslide in Fujian and Shanxi provinces, achieving an F1-score of 95.33%. Bhuyan et al. (2023) applied the U-Net model to medium-resolution RapidEye images from 2013 to 2019 in Nepal's Rasuwa region to automatically detect and map landslides over different time periods, obtaining an average F1-score of 0.69 in the study area. Fang et al. (2022) proposed a method based on LiDAR data and an improved U-Net for identifying historical landslides in densely vegetated mountainous areas. They tested the method in the Jiuzhaigou earthquake-affected region, achieving high-precision landslide identification with relatively low computational costs.

1.2 Importance of Landslide Identification and Susceptibility Mapping

1.2.1 Challenges in Landslides Identification and Susceptibility Mapping

The complex geographic environment surrounding landslides (including topography, geology, hydrology, and vegetation cover) and human activities (such as mining, road construction, and deforestation) contribute to diverse and intricate landslide characteristics, posing significant challenges for large-scale landslide identification and susceptibility mapping. Additionally, landslides often occur under adverse weather conditions, such as heavy rainfall and cloud cover (Shen, 2022). Optical remote sensing imagery is particularly susceptible to weather and lighting conditions, leading to degraded image quality. These factors make landslide identification based solely on optical remote sensing imagery difficult and limited.

SAR can image normally under harsh weather conditions and possesses certain penetration capabilities, allowing it to penetrate clouds, fog, and vegetation, enabling all-weather, day-and-night observation (Plank et al., 2016). However, SAR imagery has inherent limitations, including layover, shadow, and foreshortening effects, which result in the loss of information in certain areas. Similarly, Interferometric SAR (InSAR) technology has limitations in landslide identification and susceptibility mapping. In densely vegetated areas, InSAR technology significantly reduces the accuracy of interferometric measurements, leading to incomplete or entirely missing deformation point measurements. Moreover, rapid landslide deformation may exceed the maximum measurable range of InSAR, rendering it ineffective (Liang et al., 2022).

Existing landslide susceptibility mapping models, such as geomorphological mapping (Parry, 2011), index-based methods (Yalcin, 2008), and landslide inventory analysis (Galli et al., 2008), often rely on expert opinions and judgments, which can lack objectivity (Wang et al., 2020). In contrast, quantitative models, which analyze landslides from a numerical perspective, are widely used in landslide susceptibility mapping. Quantitative models can be further divided into deterministic models and data-driven models (Aditian et al., 2018). The quality of deterministic models largely depends on factors such as landslide type, detailed and uniform soil mechanical parameters, and hydrological environmental conditions (Pradhan & Kim, 2016). Strict data requirements often limit the applicability of these models across different study areas.

In comparison, data-driven models only require the determination of input and output data to assess landslide susceptibility over large areas. Among data-driven models, advanced machine learning methods such as Bayesian networks, decision trees, random forests, and support vector machines have demonstrated superior performance in landslide susceptibility mapping compared to heuristic and general statistical models (Song et al., 2012). However, as scenarios become increasingly complex and the demand for prediction accuracy grows, traditional machine learning methods are unable to fully extract underlying landslide features. Additionally, these models suffer from issues such as overfitting and a tendency to converge to local optima, making them increasingly inadequate for practical applications (He et al., 2021).

1.2.2 The Importance of Landslide Identification and Susceptibility Mapping for Landslide Prevention and Mitigation

First, landslide identification serves as the foundational step in landslide management and control. Traditional methods relying on field surveys are inefficient and susceptible to subjective influences. Intelligent landslide identification utilizing InSAR and deep learning technologies can accurately detect potential landslides, significantly improving identification accuracy. Second, landslide susceptibility mapping predicts the spatial probability of landslide occurrence by analyzing factors such as topography, geology, precipitation, and human activities. Deep learning-based mapping models can optimize hazard risk zoning, guiding infrastructure development away from high-risk areas. Precise evaluation results assist governments in formulating targeted mitigation measures, including slope reinforcement, drainage system construction, and ecological restoration.

1.3 Research Progress in Intelligent Landslide Identification

Deep learning, an emerging research direction in machine learning, utilizes multi-layer algorithms or neural networks to recognize and learn feature representations from data (Yang, 2024). In recent years, advances in deep learning have provided new approaches for landslide identification, gradually being applied in this field with superior results compared to traditional machine learning methods (Bragagnolo et al., 2021; Ghorbanzadeh et al., 2021; Su et al., 2021). Among these, methods combining Convolutional Neural Networks (CNN) and U-Net with optical remote sensing imagery have been widely adopted for landslide detection tasks.

Shi et al. (2021) proposed a CNN-based landslide identification method integrated with change detection using remote sensing images and validated its effectiveness in two landslide-prone areas in Hong Kong, China, achieving over 80% accuracy with high processing speed. Wang et al. (2019) employed CNN with 9-band Gaofen-1 imagery and NDVI indices to identify landslides in Fujian and Shanxi provinces, obtaining an F1-score of 95.33%. Bhuyan et al. (2023) applied the U-Net model to medium-resolution RapidEye images (2013–2019) in Nepal's Rasuwa region for automated landslide detection across different periods, achieving an average F1-score of 0.69. Fang et al. (2022) developed an improved U-Net approach using LiDAR data for historical landslide identification in densely vegetated mountainous areas, demonstrating high-precision detection with low computational costs in the Jiuzhaigou earthquake-affected region.

However, the complex geographic environments surrounding landslides (including topography, geology, hydrology, and vegetation cover) and human activities (such as mining, road construction, and deforestation) create diverse and intricate landslide characteristics. Additionally, landslides often occur under adverse weather conditions like heavy rainfall and cloud cover, which degrade optical imagery quality due to atmospheric interference. These factors limit the reliability of landslide identification relying solely on optical remote sensing data.

SAR overcomes these limitations by operating under all-weather conditions with penetration capabilities through clouds and vegetation (Plank et al., 2016). Moreover, as landslide formation closely correlates with topographic features (Dai et al., 2002), SAR's sensitivity to surface morphology makes it an ideal data source when combined with terrain data (Di Martire et al., 2016). Nava et al. (2022) demonstrated this by applying CNN to SAR data for landslide identification in Iburi Subprefecture, Hokkaido, Japan. Meena et al. (2022) evaluated U-Net and machine learning methods using RapidEye optical data, ALOS-PALSAR, and derived terrain parameters (DEM, slope) for automated landslide detection in the Himalayas.

Notably, the most current deep learning-based landslide identification methods rely on single-model architectures with unimodal remote sensing data. These models face inherent limitations in design and parameterization, typically capturing only localized features while struggling with data scarcity, noise, or anomalies—leading to overfitting and reduced robustness in complex scenarios. Furthermore, in heterogeneous environments, single-source remote sensing data cannot adequately represent the multidimensional characteristics (spectral, topographic, and textural) required for accurate landslide identification, potentially introducing biases in results.

1.4 Research Progress in Landslide Susceptibility Mapping

Landslide susceptibility mapping models can generally be classified into qualitative and quantitative models. Existing qualitative models, such as geomorphological mapping (Parry, 2011), index-based methods (Yalcin, 2008), and landslide inventory analysis (Galli et al., 2008), often rely on expert opinions and judgments, which may lack objectivity (Wang et al., 2020). In contrast, quantitative models analyze landslide occurrence from a numerical perspective and are widely used in landslide susceptibility mapping.

Quantitative models can be further divided into deterministic models and data-driven models (Aditian et al., 2018). The quality of deterministic models largely depends on factors such as landslide type, detailed and uniform soil mechanical parameters, and hydrological environmental conditions (Pradhan & Kim, 2016). However, strict data requirements often limit their applicability across different study areas. In comparison, data-driven models only require input and output data to assess landslide susceptibility over large regions. Among data-driven models, advanced machine learning methods such as Bayesian networks, decision trees, random forests, and support vector machines have demonstrated superior performance over heuristic and general statistical models in landslide susceptibility assessment (Song et al., 2012). Nevertheless, as scenarios become increasingly complex and the demand for prediction accuracy grows, traditional machine learning methods struggle to fully extract underlying landslide features and often suffer from issues such as overfitting and local optima, making them inadequate for practical applications (He et al., 2021).

Deep learning models effectively overcome the limitations of traditional machine learning models. Their deep architectures can efficiently identify optimal features and handle indirect relationships between features and targets, thereby simplifying feature engineering and data preprocessing steps. In landslide susceptibility mapping, Song et al. (2012) first proposed optimized CNN architectures based on LeNet-5, which significantly outperformed traditional machine learning methods in terms of overall accuracy and Matthews correlation coefficient, demonstrating the potential of CNNs in landslide susceptibility mapping. Subsequently, Tang (2019) constructed an AlexNet-2 CNN model for landslide susceptibility mapping and found that CNNs achieved higher accuracy than traditional machine learning models, further confirming their feasibility.

Another widely used deep learning model in landslide susceptibility mapping is the Recurrent Neural Network (RNN), which employs internal memory units to process sequential inputs. As a complex evolutionary process, landslide formation and occurrence are essentially responses to the interactions of multiple unstable factors. The recursive structure of RNNs helps retain the most critical information related to landslide occurrence and propagate it to subsequent hidden states. Wang et al. (2020) ranked each landslide-influencing factor layer in descending order of importance. By treating these factors as single-band images and converting each pixel into sequential samples based on importance, they achieved satisfactory results.

In recent developments, researchers have increasingly focused on improving accuracy and computational efficiency through algorithm optimization and model integration. For example, He et al. (2021) combined 2D CNNs with Gated Recurrent Units (GRUs) to extract neighborhood and sequential features of landslides for comprehensive characterization. Lv et al. (2022) proposed a heterogeneous ensemble learning method incorporating CNNs and other deep learning models, achieving results superior to those of individual models. However, current models in landslide susceptibility mapping primarily utilize static features of influencing factors, with limited research on dynamic data such as InSAR deformation characteristics.

Some scholars have attempted to integrate InSAR results with landslide susceptibility mapping. For instance, He et al. (2023) combined MLP-based susceptibility mapping results with InSAR identification results to design a threshold model for high-risk landslide zoning. Yuan and Chen (2022) used InSAR data to screen landslides and investigated the impact of InSAR identification results and overall deformation on susceptibility mapping. Their findings suggest that preliminary InSAR identification and incorporating InSAR deformation results as input factors can improve susceptibility mapping, demonstrating potential value (Gao et al., 2023).

1.5 Summary

This chapter presents the global economic and human impacts of landslides, including their hazard scope and distribution patterns. It specifically examines landslide effects in China and discusses the limitations of traditional landslide identification and susceptibility mapping methods. The challenges in landslide detection and susceptibility evaluation are systematically analyzed. Furthermore, it reviews the developmental progress in intelligent landslide identification and susceptibility mapping technologies, while critically evaluating the advantages and disadvantages of various technical approaches.

References

Aditian, A., Kubota, T., & Shinohara, Y. (2018). Comparison of GIS-based landslide susceptibility models using frequency ratio, logistic regression, and artificial neural network in a tertiary region of Ambon, Indonesia. *Geomorphology, 318*, 101–111. https://doi.org/10.1016/j.geomorph.2018.06.006

Agwe, J. N., Arnold, M., Buys, P., Chen, R. S., Deichmann, U. K., Dilley, M., Yetman, G. (2005). Natural disaster hotspots: A global risk analysis. World Bank. http://documents.worldbank.org/curated/en/621711468175150317

Alcántara-Ayala, I. (2002). Geomorphology, natural hazards, vulnerability and prevention of natural disasters in developing countries. *Geomorphology, 47*(2–4), 107–124. https://doi.org/10.1016/S0169-555X(02)00083-1

Bank, W. (2014). Global Facility for Disaster Reduction and Recovery. World Bank. https://hdl.handle.net/10986/20682

Berardino, P., Fornaro, G., Lanari, R., & Sansosti, E. (2002). A new algorithm for surface deformation monitoring based on small baseline differential SAR interferograms. *IEEE Transactions on Geoscience and Remote Sensing, 40*(11), 2375–2383. https://doi.org/10.1109/TGRS.2002.803792

Bhuyan, K., Meena, S. R., Nava, L., van Westen, C., Floris, M., Catani, F. (2023). Mapping landslides through a temporal lens: An insight toward multi-temporal landslide mapping using the U-Net deep learning model. *GIScience & Remote Sensing*, 60(1), Article 2182057. https://doi.org/10.1080/15481603.2023.2182057

Blaschke, T., Feizizadeh, B., & Höfling, D. (2014). Object-based image analysis and digital terrain analysis for locating landslides in the Urmia Lake Basin, Iran. *IEEE Journal of Selected Topics in Applied Earth Observations and Remote Sensing, 7*(12), 4806–4817. https://doi.org/10.1109/JSTARS.2014.2350036

Bragagnolo, L., Rezende, L. R., da Silva, R. V., Grzybowski, J. M. V. (2021). Convolutional neural networks applied to semantic segmentation of landslide scars. *Catena*, 201* Article 105189. https://doi.org/10.1016/j.catena.2021.105189

Chen, T., Trinder, J. C., & Niu, R. (2017). Object-oriented landslide mapping using ZY-3 satellite imagery, random forest and mathematical morphology, for the Three-Gorges Reservoir, China. *Remote Sensing*, 9(4), Article 333. https://doi.org/10.3390/rs9040333

Chen, Y., Song, C., Li, Z., Yu, C., Liu, Z., Zhang, X., Hu, X. (2024). Characterizing the evolution of the Daguangbao landslide nearly 15 years after the 2008 Wenchuan earthquake by InSAR observations. *Engineering Geology*, 342, Article 107748. https://doi.org/10.1016/j.enggeo.2024.107748

Chen, H., He, Y., Zhang, L., Yang, W., Liu, Y., Gao, B., & Lu, J. (2024a). A multi-input channel U-Net landslide detection method fusing SAR multisource remote sensing data. *IEEE Journal of Selected Topics in Applied Earth Observations and Remote Sensing, 17*, 1215–1232. https://doi.org/10.1109/JSTARS.2023.3339294

Choi, K. Y., & Cheung, R. W. M. (2013). Landslide disaster prevention and mitigation through works in Hong Kong. *Journal of Rock Mechanics and Geotechnical Engineering, 5*(5), 354–365. https://doi.org/10.1016/j.jrmge.2013.07.007

Cruden, D. M. (1991). A simple definition of a landslide. *Bulletin of the International Association of Engineering Geology, 43*(1), 27–29. https://doi.org/10.1007/BF02590167

Dai, F. C., Lee, C. F., & Ngai, Y. Y. (2002). Landslide risk assessment and management: An overview. *Engineering Geology, 64*(1), 65–87. https://doi.org/10.1016/S0013-7952(01)00093-X

Di Martire, D., Tessitore, S., Brancato, D., Ciminelli, M. G., Costabile, S., Costantini, M., & Calcaterra, D. (2016). Landslide detection integrated system (LaDIS) based on in-situ and satellite SAR interferometry measurements. *CATENA, 137*, 406–421. https://doi.org/10.1016/j.catena.2015.10.002

Fang, C., Fan, X., Zhong, H., Lombardo, L., Tanyas, H., Wang, X. (2022). A novel historical landslide detection approach based on LiDAR and lightweight attention U-Net. *Remote Sensing*, 14(17), Article 4357. https://doi.org/10.3390/rs14174357

Froude, M. J., & Petley, D. N. (2018). Global fatal landslide occurrence from 2004 to 2016. *Natural Hazards and Earth System Sciences, 18*(8), 2161–2181. https://doi.org/10.5194/nhess-18-2161-2018

Galli, M., Ardizzone, F., Cardinali, M., Guzzetti, F., & Reichenbach, P. (2008). Comparing landslide inventory maps. *Geomorphology, 94*(3–4), 268–289. https://doi.org/10.1016/j.geomorph.2006.09.023

Gao, B. H., He, Y., Zhang, L. F., Yao, S., & Yang, W. (2023). Dynamic assessment of landslide susceptibility considering InSAR deformation: A case study of the Liujiaxia Reservoir area. *Chinese Journal of Rock Mechanics and Engineering, 42*(2), 450–465. [in Chinese].

Ghorbanzadeh, O., Blaschke, T., Gholamnia, K., Meena, S. R., Tiede, D., & Aryal, J. (2019). Evaluation of different machine learning methods and deep-learning convolutional neural networks for landslide detection. *Remote Sensing, 11*(2), 196. https://doi.org/10.3390/rs11020196

Ghorbanzadeh, O., Crivellari, A., Ghamisi, P., Shahabi, H., & Blaschke, T. (2021). A comprehensive transferability evaluation of U-Net and ResU-Net for landslide detection from Sentinel-2 data (case study areas from Taiwan, China, and Japan). *Scientific Reports, 11*(1), 14629. https://doi.org/10.1038/s41598-021-94190-9

Ghorbanzadeh, O., Shahabi, H., Crivellari, A., Homayouni, S., Blaschke, T., & Ghamisi, P. (2022). Landslide detection using deep learning and object-based image analysis. *Landslides, 19*(4), 929–939. https://doi.org/10.1007/s10346-021-01843-x

Gu, D. M., Huang, D., Yang, W. D., Zhu, J. L., & Fu, G. Y. (2017). Understanding the triggering mechanism and possible kinematic evolution of a reactivated landslide in the Three Gorges Reservoir. *Landslides, 14*(6), 2073–2087. https://doi.org/10.1007/s10346-017-0845-4

He, Y., Zhao, Z. a., Yang, W., Yan, H., Wang, W., Yao, S., Liu, T. (2021). A unified network of information considering superimposed landslide factors sequence and pixel spatial neighbourhood for landslide susceptibility mapping. *International Journal of Applied Earth Observation and Geoinformation*, 104, Article 102508. https://doi.org/10.1016/j.jag.2021.102508

He, Y., Wang, W., Zhang, L., Chen, Y., Chen, Y., Chen, B., Zhao, Z. (2023). An identification method of potential landslide zones using InSAR data and landslide susceptibility. Geomatics, Natural Hazards and Risk, 14(1), Article 2185120. https://doi.org/10.1080/19475705.2023.2185120

He, Y., Huo, T., Gao, B., Zhu, Q., Jin, L., Chen, J., & Tang, J. (2024). Thaw slump susceptibility mapping based on sample optimization and ensemble learning techniques in Qinghai-Tibet railway corridor. *IEEE Journal of Selected Topics in Applied Earth Observations and Remote Sensing, 17*, 5443–5459. https://doi.org/10.1109/JSTARS.2024.3368039

He, Y., Yan, H., Yang, W., Yao, S., Zhang, L., Chen, Y., & Liu, T. (2022). Time-series analysis and prediction of surface deformation in the Jinchuan mining area, Gansu province, by using InSAR and CNN–PhLSTM network. *IEEE Journal of Selected Topics in Applied Earth Observations and Remote Sensing, 15*, 6732–6751. https://doi.org/10.1109/JSTARS.2022.3198728

Herath, S., & Wang, Y. (2009). *Case studies and national experiences.* Springer.

Here is your reference list formatted in **APA 7th edition** style, with corrections for consistency, proper formatting of author names, journal titles, and DOI/URL links.

Highland, L. M., & Bobrowsky, P. (2008). The landslide handbook—A guide to understanding landslides (U.S. Geological Survey Circular No. 1325). U.S. Geological Survey.

Huang, F., Yao, C., Liu, W., Li, Y., & Liu, X. (2018). Landslide susceptibility assessment in the Nantian area of China: A comparison of frequency ratio model and support vector machine. *Geomatics, Natural Hazards and Risk, 9*(1), 919–938. https://doi.org/10.1080/19475705.2018.1482963

Huang, Y., Zhang, J., He, H., Jia, Y., Chen, R., Ge, Y., & Li, H. (2024). MAST: An earthquake-triggered landslides extraction method combining morphological analysis edge recognition with Swin-Transformer deep learning model. *IEEE Journal of Selected Topics in Applied Earth Observations and Remote Sensing, 17*, 2586–2595. https://doi.org/10.1109/JSTARS.2023.3342989

Jiang, W. D. (2023). Research on remote sensing intelligent identification of landslide disasters based on multi-source data [Master's thesis, Chang'an University]. CNKI.

Keyport, R. N., Oommen, T., Martha, T. R., Sajinkumar, K. S., & Gierke, J. S. (2018). A comparative analysis of pixel-and object-based detection of landslides from very high-resolution images. *International journal of applied earth observation and geoinformation, 64*, 1–11.

Kjekstad, O., & Highland, L. (2009). Economic and social impacts of landslides. In *Landslides—disaster risk reduction* (pp. 573–587). Springer. https://doi.org/10.1007/978-3-540-69970-5_30

Li, H., He, Y., Xu, Q., Deng, J., Li, W., & Wei, Y. (2022). Detection and segmentation of loess landslides via satellite images: A two-phase framework. *Landslides, 19*(3), 673–686. https://doi.org/10.1007/s10346-021-01789-0

Li, X., Li, Z., Yang, J., Liu, Y., Fu, B., Qi, W., & Fan, X. (2018). Spatiotemporal characteristics of earthquake disaster losses in China from 1993 to 2016. *Natural Hazards, 94*(2), 843–865. https://doi.org/10.1007/s11069-018-3425-6

Li, Z., Shi, W., Lu, P., Yan, L., Wang, Q., & Miao, Z. (2016). Landslide mapping from aerial photographs using change detection-based Markov random field. *Remote Sensing of Environment, 187*, 76–90. https://doi.org/10.1016/j.rse.2016.10.008

Liang, J., Dong, J., Zhang, S., Zhao, C., Liu, B., Yang, L., & Ma, X. (2022). Discussion on InSAR identification effectivity of potential landslides and factors that influence the effectivity. *Remote Sensing, 14*(8), 1952. https://doi.org/10.3390/rs14081952

Liao, M., Tang, J., Wang, T., Balz, T., & Zhang, L. (2011). Landslide monitoring with high-resolution SAR data in the Three Gorges region. *Science China Earth Sciences, 55*(4), 590–601. https://doi.org/10.1007/s11430-011-4259-1

Lin, Q., & Wang, Y. (2018). Spatial and temporal analysis of a fatal landslide inventory in China from 1950 to 2016. *Landslides, 15*(12), 2357–2372. https://doi.org/10.1007/s10346-018-1037-6

Lv, L., Chen, T., Dou, J., & Plaza, A. (2022). A hybrid ensemble-based deep-learning framework for landslide susceptibility mapping. *International Journal of Applied Earth Observation and Geoinformation, 108*, Article 102713. https://doi.org/10.1016/j.jag.2022.102713

Meena, S. R., Soares, L. P., Grohmann, C. H., van Westen, C., Bhuyan, K., Singh, R. P., & Catani, F. (2022). Landslide detection in the Himalayas using machine learning algorithms and U-Net. *Landslides, 19*(5), 1209–1229. https://doi.org/10.1007/s10346-022-01861-3

Nava, L., Monserrat, O., & Catani, F. (2022). Improving landslide detection on SAR data through deep learning. *IEEE Geoscience and Remote Sensing Letters, 19*, 1–5. https://doi.org/10.1109/LGRS.2021.3127073

Parry, S. (2011). The application of geomorphological mapping in the assessment of landslide hazard in Hong Kong. In *Developments in Earth Surface Processes* (Vol. 15, pp. 413–441). Elsevier. https://doi.org/10.1016/B978-0-444-53446-0.00016-5

Pascaline, W., & Rowena, H. (2019). Economic losses, poverty & disasters: 1998–2017. Centre for Research on the Epidemiology of Disasters (CRED).

Petley, D. (2011). Global deaths from landslides in 2010 (updated to include a comparison with previous years) [Blog post]. *The Landslide Blog*. https://blogs.agu.org/landslideblog/2011/02/05/global-deaths-from-landslides-in-2010/

Petschko, H., Bell, R., & Glade, T. (2016). Effectiveness of visually analyzing LiDAR DTM derivatives for earth and debris slide inventory mapping for statistical susceptibility modeling. *Landslides, 13*(5), 857–872. https://doi.org/10.1007/s10346-015-0622-1

Plank, S., Twele, A., & Martinis, S. (2016). Landslide mapping in vegetated areas using change detection based on optical and polarimetric SAR data. *Remote Sensing, 8*(4), 307. https://doi.org/10.3390/rs8040307

Pradhan, A. M. S., & Kim, Y. T. (2016). Evaluation of a combined spatial multi-criteria evaluation model and deterministic model for landslide susceptibility mapping. *CATENA, 140*, 125–139. https://doi.org/10.1016/j.catena.2016.01.022

Prakash, N., Manconi, A., & Loew, S. (2021). A new strategy to map landslides with a generalized convolutional neural network. *Scientific Reports, 11*(1), 9722. https://doi.org/10.1038/s41598-021-89015-8

Redshaw, P., & Bottomley, J. (2020). *The world bank—the global landslide hazard map—final project report*. World Bank.

Rosi, A., Tofani, V., Tanteri, L., Tacconi Stefanelli, C., Agostini, A., Catani, F., & Casagli, N. (2017). The new landslide inventory of Tuscany (Italy) updated with PS-InSAR: Geomorphological features and landslide distribution. *Landslides, 15*(1), 5–19. https://doi.org/10.1007/s10346-017-0861-4

Shanmugam, G., & Wang, Y. (2015). The landslide problem. *Journal of Palaeogeography, 4*(2), 109–166. https://doi.org/10.3724/SP.J.1261.2015.00071

Shen, Q. (2022). Extraction and spatiotemporal characteristics analysis of rainfall-induced landslide events based on multi-source remote sensing data [Master's thesis, Nanjing Normal University]. CNKI.

Shi, W., Zhang, M., Ke, H., Fang, X., Zhan, Z., & Chen, S. (2021). Landslide recognition by deep convolutional neural network and change detection. *IEEE Transactions on Geoscience and Remote Sensing, 59*(6), 4654–4672. https://doi.org/10.1109/TGRS.2020.3015826

Song, Y., Gong, J., Gao, S., Wang, D., Cui, T., Li, Y., & Wei, B. (2012). Susceptibility assessment of earthquake-induced landslides using Bayesian network: A case study in Beichuan, China. *Computers & Geosciences, 42*, 189–199. https://doi.org/10.1016/j.cageo.2011.09.011

Su, Z., Chow, J. K., Tan, P. S., Wu, J., Ho, Y. K., & Wang, Y.-H. (2021). Deep convolutional neural network–based pixel-wise landslide inventory mapping. *Landslides, 18*(4), 1421–1443. https://doi.org/10.1007/s10346-020-01557-6

Tang, X. N. (2019). Landslide susceptibility assessment in Lvliang City based on convolutional neural network and comprehensive index model [Master's thesis, Taiyuan University of Technology].

Tien Bui, D., Hoang, N.-D., Martínez-Álvarez, F., Ngo, P.-T.T., Hoa, P. V., Pham, T. D., & Costache, R. (2020). A novel deep learning neural network approach for predicting flash flood susceptibility: A case study at a high frequency tropical storm area. *Science of the Total Environment, 701*, Article 134413. https://doi.org/10.1016/j.scitotenv.2019.134413

Wang, X. L., Hu, D. Y., Tang, H., & Shu, Y. (2014). Landslide information extraction from airborne fully polarimetric SAR images based on Bayes decision. *Remote Sensing for Land & Resources, 26*(2), 121–127. [in Chinese].

Wang, Y., Fang, Z., Wang, M., Peng, L., & Hong, H. (2020). Comparative study of landslide susceptibility mapping with different recurrent neural networks. *Computers & Geosciences, 138*, Article 104445. https://doi.org/10.1016/j.cageo.2020.104445

Wang, Y., Wang, X., & Jian, J. (2019). Remote sensing landslide recognition based on convolutional neural network. *Mathematical Problems in Engineering, 2019*, 8389368. https://doi.org/10.1155/2019/8389368

Wu, L., Liu, R., Ju, N., Zhang, A., Gou, J., He, G., & Lei, Y. (2024). Landslide mapping based on a hybrid CNN-transformer network and deep transfer learning using remote sensing images with topographic and spectral features. *International Journal of Applied Earth Observation and Geoinformation, 126*, Article 103612. https://doi.org/10.1016/j.jag.2023.103612

Wu, M. Y., Luo, M., & Liu, S. H. (2022). Integrated identification of potential landslides and old landslides based on optical remote sensing and InSAR technology: A case study in northwestern Yunnan. *The Chinese Journal of Geological Hazard and Control, 33*(4), 84–93. [in Chinese].

Xiao, H., Huang, J., Ma, Q., Wan, J., Li, L., Peng, Q., & Rezaeimalek, S. (2015). Experimental study on the soil mixture to promote vegetation for slope protection and landslide prevention. *Landslides, 14*(1), 287–297. https://doi.org/10.1007/s10346-015-0634-x

Xu, C., Dai, F. C., Chen, J., Tu, X. B., Xu, L., Li, W. C., Yao, X. (2009). Detailed interpretation of secondary geological disasters in the Wenchuan Ms8.0 earthquake-affected area using remote sensing. *Journal of Remote Sensing*, 13(4), 754–762. [in Chinese]

Xu, C., Xu, X., Shyu, J. B. H., Gao, M., Tan, X., Ran, Y., & Zheng, W. (2015). Landslides triggered by the 20 April 2013 Lushan, China, Mw 6.6 earthquake from field investigations and preliminary analyses. *Landslides*, 12(2), 365–385. https://doi.org/10.1007/s10346-014-0546-1

Xu, Q., Zhao, B., Dai, K., Dong, X., Li, W., Zhu, X., & Ge, D. (2023). Remote sensing for landslide investigations: A progress report from China. *Engineering Geology, 321*, Article 107156. https://doi.org/10.1016/j.enggeo.2023.107156

Yalcin, A. (2008). GIS-based landslide susceptibility mapping using analytical hierarchy process and bivariate statistics in Ardesen (Turkey): Comparisons of results and confirmations. *CATENA, 72*(1), 1–12. https://doi.org/10.1016/j.catena.2007.01.003

Yang, W. (2024). Key InSAR data processing methods for high-precision deformation monitoring using deep neural networks [Master's thesis, Lanzhou Jiaotong University]. CNKI.

Yang, W., Wang, M., & Shi, P. (2013). Using MODIS NDVI time series to identify geographic patterns of landslides in vegetated regions. *IEEE Geoscience and Remote Sensing Letters, 10*(4), 707–710. https://doi.org/10.1109/LGRS.2012.2219576

Yuan, R., & Chen, J. (2022). A hybrid deep learning method for landslide susceptibility analysis with the application of InSAR data. *Natural Hazards, 114*(2), 1393–1426. https://doi.org/10.1007/s11069-022-05430-8

Zhang, Y. (2018). Research on surface deformation monitoring and early landslide identification based on InSAR technology [Doctoral dissertation]. CNKI. https://kns.cnki.net/kcms2/article/abstract?v=9hl5eXOdJcYmD_nDBKE3nafIsvD9Pyy5XLosMh1PxnrhhuBbrxJFevIXJ1qDk5SSejaviR-sKSEKN2p9M6lWAZWU8xTA-n-8tHKT8yFEGLBoyp_3UffllRvYD-HqkaGYjGW6o1CihAPtmuex43xOXQ==

Zhang, Q., Zhao, C. Y., & Chen, X. R. (2022). Advances and development trends in multi-source remote sensing technology for early identification of geological hazards. *Acta Geodaetica Et Cartographica Sinica, 51*(6), 885–896. [in Chinese].

Zhao, Z. A. (2024). An integrated dynamic-static feature landslide susceptibility assessment method incorporating InSAR deformation information [Master's thesis, Lanzhou Jiaotong University]. CNKI.

Zhao, W., Li, A., Nan, X., Zhang, Z., & Lei, G. (2017). Postearthquake landslides mapping from Landsat-8 data for the 2015 Nepal earthquake using a pixel-based change detection method. *IEEE Journal of Selected Topics in Applied Earth Observations and Remote Sensing, 10*(5), 1758–1768. https://doi.org/10.1109/JSTARS.2017.2661802

Zhou, H. J., Wang, X., & Yuan, Y. (2015). Risk assessment of disaster chain: Experience from Wenchuan earthquake-induced landslides in China. *Journal of Mountain Science, 12*(5), 1169–1180. https://doi.org/10.1007/s11629-015-3453-6

Chapter 2
Theories of InSAR and Deep Learning

2.1 Basic Principles of InSAR

InSAR is a technique that utilizes two or more SAR images to perform phase differential processing, thereby extracting information about terrain, landforms, or ground surface deformation. The concept of interferometry originates from Young's double-slit interference experiment (Feynman, 1963). As shown in Fig. 2.1, when a light source shines on two slits on a plate, alternating bright and dark interference fringes form on the receiving screen behind the slits. This phenomenon is referred to as optical interference. Similar to Young's double-slit experiment, the two satellite observation positions in InSAR correspond to the two slits on the plate, the interference fringes on the receiving screen correspond to the InSAR interferometric phase, and the receiving screen represents the observed ground surface in InSAR. When ground surface deformation or ground elevation changes occur, the InSAR interferometric phase exhibits variations related to the deformation or elevation changes. By leveraging this information, we can invert ground surface deformation or topographic data (Liu, 2019).

This chapter will focus on the fundamental principles of InSAR and the basic principles of Differential InSAR (D-InSAR) ground surface deformation monitoring technology. It will also provide a detailed explanation of the fundamental principles and application cases of time-series InSAR deformation monitoring techniques, including two classic approaches: Small Baseline Subset InSAR (SBAS-InSAR) and Persistent Scatterer InSAR (PS-InSAR).

2.1.1 InSAR Introduction

InSAR technology originated from Young's double-slit interference experiment (Feynman, 1963) and combines SAR imaging with interferometric measurement

Y. He, *InSAR and Deep Learning in Landslides Research: Intelligent Identification, Risk Assessment and Susceptibility Mapping*, https://doi.org/10.1007/978-981-96-9132-6_2

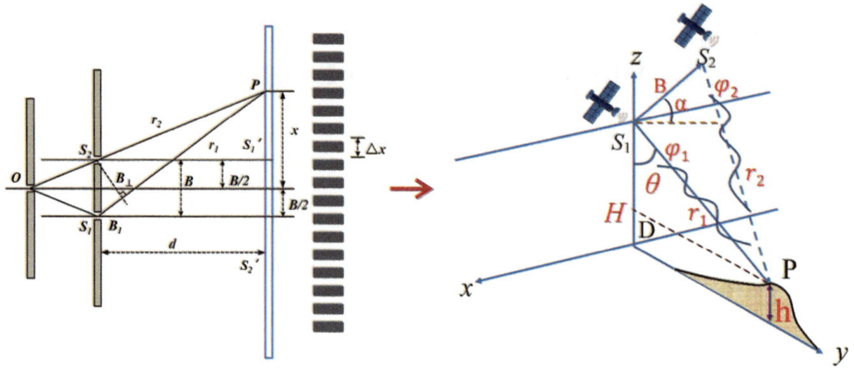

(a) 杨氏双缝干涉实验 (b) InSAR观测几何

Fig. 2.1 Young's double-slit interference experiment and InSAR observation geometry

techniques (Wright et al., 2004). By utilizing two SAR images of the same area and analyzing the phase differences of corresponding points based on radar sensor parameters and imaging geometry, it retrieves ground elevation data. With advancements in radio interferometry and spaceborne radar satellites, radar remote sensing data have been increasingly applied in fields such as mapping, oceanography, and hazards prevention (Jordan, 1980). Due to its stable and high-precision observation capabilities, InSAR has become an indispensable tool for monitoring ground surface deformation and geological hazards.

Figure 2.2 illustrates the interferometric geometry of InSAR. Here, S_1 and S_2 represent the positions of the radar sensor at two different imaging times over target point P, where S_1 corresponds to the master image and S_2 to the slave image. The distance between S_1 and S_2 is called the spatial baseline (B), while R_1 and R_2 denote the slant ranges from S_1 and S_2 to P, respectively. The altitude (H) of S_1 is its vertical distance above the Earth's reference ellipsoid. The angle α represents the tilt of the baseline B relative to the horizontal plane, and θ is the incidence angle of S_1.

The spatial baseline (B) can be decomposed into two orthogonal components, perpendicular baseline ($B_\perp$): Normal to the radar line of sight (LOS). Parallel baseline ($B_\parallel$): Along the LOS direction. Their calculation follows (Hanssen, 2001):

$$B_\perp = B\cos(\theta - \alpha) \tag{2.1}$$

$$B_\parallel = B\sin(\theta - \alpha) \tag{2.2}$$

Since the reference ellipsoid can be approximated as a plane over small areas, the elevation (h) of target point P can be derived from the geometric relationship shown in Fig. 2.2. The key parameters include: H: Platform height of the master satellite (S_1) above the reference ellipsoid.

R_1: Slant range distance from S_1 to P.

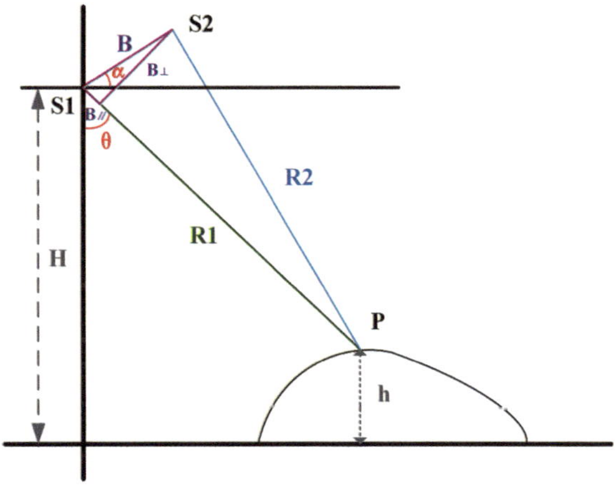

Fig. 2.2 InSAR interferometric imaging geometry schematic

θ: Incidence angle of the radar signal.

The elevation (h) is calculated as:

$$h = H - R_1\cos\theta \tag{2.3}$$

Since H and the incidence angle θ can be determined from the sensor's position and parameters, the value of R_1 must be calculated to solve for h.

In the triangle S_1PS_2, the law of cosines gives:

$$\cos(90^\circ - \theta + \alpha) = \frac{{R_1}^2 - {R_2}^2 + B^2}{2R_1B} \tag{2.4}$$

Let $\Delta R = R_2 - R_1$, then the phase difference ($\Delta\varphi$) in the interferometric signal propagation can be expressed as:

$$\Delta\varphi = \varphi_1 - \varphi_2 = -\frac{4\pi}{\lambda}\Delta R \tag{2.5}$$

Here, λ represents the known parameter of the radar wavelength. The interferometric phase difference $\Delta\varphi$ can be obtained by conjugate multiplication of the two single-look complex (SLC) radar images acquired at positions S_1 and S_2. However, $\Delta\varphi$ is initially wrapped within the range of $(-\pi, \pi]$. To accurately derive the topographic information of the observed target P, the true phase value must be obtained through phase unwrapping (Chen et al., 2012).

In the triangle ΔS_1PS_2, since $R_1 \gg B$ and $R_1 \gg \Delta R$, combining Eqs. (2.4) and (2.5) yields an expression for the incidence angle θ:

$$\sin(\theta - \alpha) \approx \frac{\Delta R}{B} = -\frac{\Delta\varphi\lambda}{4\pi B} \tag{2.6}$$

$$\theta = \alpha - \arcsin\left(\frac{\Delta\varphi\lambda}{4\pi B}\right) \tag{2.7}$$

By combining Eqs. (2.3) and (2.7), the elevation h of the target position P can be determined.

2.1.2 Ground Surface Deformation Monitoring of D-InSAR

The fundamental concept of InSAR traces back to Thomas Young's "double-slit interference experiment" in 1801. When electromagnetic waves (or light waves) pass through two narrow slits, the difference in propagation distances to the same point on the receiving screen creates a phase difference between the two waves. This results in constructive or destructive interference, producing alternating bright and dark interference fringes on the screen (Baran et al., 2005). InSAR builds upon this principle and has evolved into a cutting-edge technology. Its core methodology involves.

Acquiring two SAR images of the same area via satellite radar along repeat orbits. Generating an interferogram by phase interference processing. Inverting 3D topographic changes or constructing a DEM by combining the interferometric phase with radar parameters and satellite positional data.

D-InSAR technology further advances InSAR. In differential interferometry, the interferometric phase (ϕ) of each pixel in the differential interferogram is composed of multiple components:

$$\phi = \phi_{\text{flat}} + \phi_{\text{top}} + \phi_{\text{def}} + \phi_{\text{atm}} + \phi_{\text{noi}} \tag{2.8}$$

In Eq. (2.8), ϕ_{flat} represents the flat-Earth phase, ϕ_{top} the topographic phase, ϕ_{def} the deformation phase, ϕ_{atm} the atmospheric phase, and ϕ_{noi} the noise phase. During the interferometric process, ϕ_{flat}, ϕ_{top}, and ϕ_{def} dominate the interferometric phase. Flat-Earth Phase (ϕ_{flat}) is caused by the reference ellipsoid, it is the primary component in the initial interferometric stage. Its value can be determined by calculating the reference ellipsoid phase and establishing a phase model. Topographic Phase (ϕ_{top}) is the results from the elevation of ground targets during SAR observations. Its magnitude depends on the perpendicular baseline—the larger the baseline, the greater the absolute value of ϕ_{top}. Deformation Phase (ϕ_{def}) captures one-dimensional displacement along the radar line of sight (LOS) if ground deformation occurs between two SAR acquisitions. This is the key information extracted in InSAR-based deformation monitoring. Atmospheric phase (ϕ_{atm}) is an unavoidable low-frequency artifact correlated with terrain and atmospheric conditions (e.g., water vapor). Noise Phase (ϕ_{noi}) is the high-frequency random noise, removable via low-pass filtering.

D-InSAR detects subtle surface deformation by acquiring SAR phase data from two distinct positions over the same area. Removing topographic phase (ϕ_{top}) using an

Fig. 2.3 Schematic diagram of D-InSAR technology

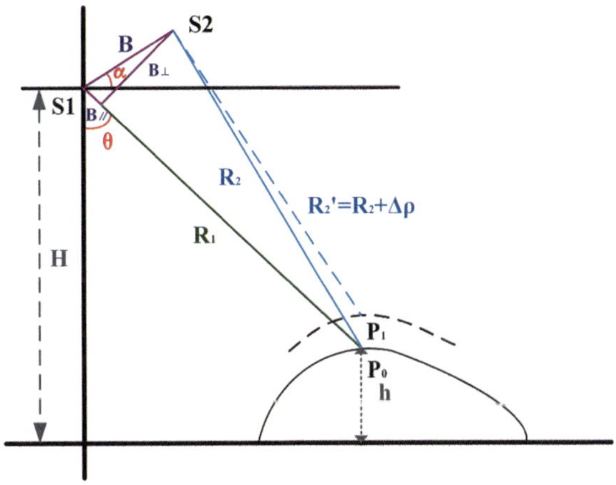

external high-precision DEM or repeat-pass interferometry. Isolating the deformation phase (ϕ_{def}) after eliminating ϕ_{flat}, ϕ_{atm}, and ϕ_{noi}. The main D-InSAR methods are two-pass method, three-pass method and four-pass method (All methods share the same core principle but differ in data usage.). Figure 2.3 illustrates the workflow of D-InSAR technology.

As shown in Fig. 2.3, during the observation of ground target $P0$, if deformation occurs at $P0$ (moving to $P1$) between the SAR acquisitions at positions $S1$ and $S2$, we define:

$R1$: Slant range from $S1$ to $P0$

$R2$: Slant range from $S2$ to $P0$

$R2\prime$: Slant range from $S2$ to $P1$ after deformation.

$\Delta\rho$: Line-of-sight (LOS) displacement (positive for movement toward the satellite). The geometric relationship satisfies:

$$\Delta\rho = R_{2}\prime - R_{2} \tag{2.9}$$

As shown in Eq. (2.5), the phase variation $\Delta\varphi$ of ground target $P0$ can be expressed as:

$$\Delta\varphi = -\frac{4\pi}{\lambda}(R_{1} - R_{2}\prime) \tag{2.10}$$

Combined Eqs. (2.9) and (2.10):

$$\Delta\varphi = -\frac{4\pi}{\lambda}(R_{1} - R_{2}) + \frac{4\pi}{\lambda}\Delta\rho \tag{2.11}$$

In Eq. (2.11), the first term comprises both ϕ_{flat} and ϕ_{flat} components, which can be eliminated using a reference ellipsoid model and external DEM data. The second term represents the deformation phase value ϕ_{def} of point $P0$, expressed as:

$$\phi_{\text{def}} = \frac{4\pi}{\lambda}\Delta\rho \tag{2.12}$$

The derived $\Delta\rho$ after transposition represents the surface deformation value at point $P0$ along the radar line-of-sight (LOS) direction from position $S2$:

$$\Delta\rho = \frac{\lambda}{4\pi}\phi_{\text{def}} \tag{2.13}$$

From Eq. (2.13), it can be seen that when ϕ_{def} completes one interference fringe (i.e., the phase change reaches 2π), the corresponding deformation equals $\lambda/2$. This value is called the deformation ambiguity of this wavelength.

This indicates that InSAR technology's sensitivity to surface deformation depends solely on the radar's wavelength. During interferometric processing, if interfering factors (such as orbit errors and atmospheric delays) can be suppressed to obtain highly coherent interferograms, the measurement of deformation along the radar line-of-sight can achieve extremely high precision.

D-InSAR technology has been widely applied in surface deformation monitoring (Howard et al., 2022), earthquakes (Liu et al., 2021; Kothyari et al., 2022; Kumar et al., 2022), landslides (Cai et al., 2020; Brasca Merlín et al., 2021; Yao et al., 2021; Niraj et al., 2022), volcanic activity (Vélez et al., 2021), and other deformation monitoring fields. Figure 2.4 illustrates the basic workflow of D-InSAR technology. When the topographic phase is removed using external DEM data, it is referred to as the two-pass method. When the topographic phase is eliminated using interferometric pairs generated from three or four SAR images, it is called the three-pass or four-pass method, respectively.

Therefore, D-InSAR technology represents an extension and advancement of InSAR in practical applications. Its fundamental principle involves performing differential interferometric processing on two or more SAR images covering the same study area to retrieve surface deformation information through inversion.

2.1.3 Time-Series InSAR Based Ground Surface Deformation Monitor Methods

The fundamental concept of TS-InSAR (time-series InSAR) technology involves utilizing phase differences from multiple images to calculate surface deformation. As illustrated in Fig. 2.5, in TS-InSAR, a time-series of images (T_1, T_2, ..., T_n) is selected, where the phase records the distances (R_1, R_2, ..., R_n) from the imaging positions to the ground observation points from the first to the nth acquisition. By computing the microwave phase differences between two observations, information about changes in surface height can be obtained. If the time interval between two observations is sufficiently short, this change can be considered linear. Therefore, by

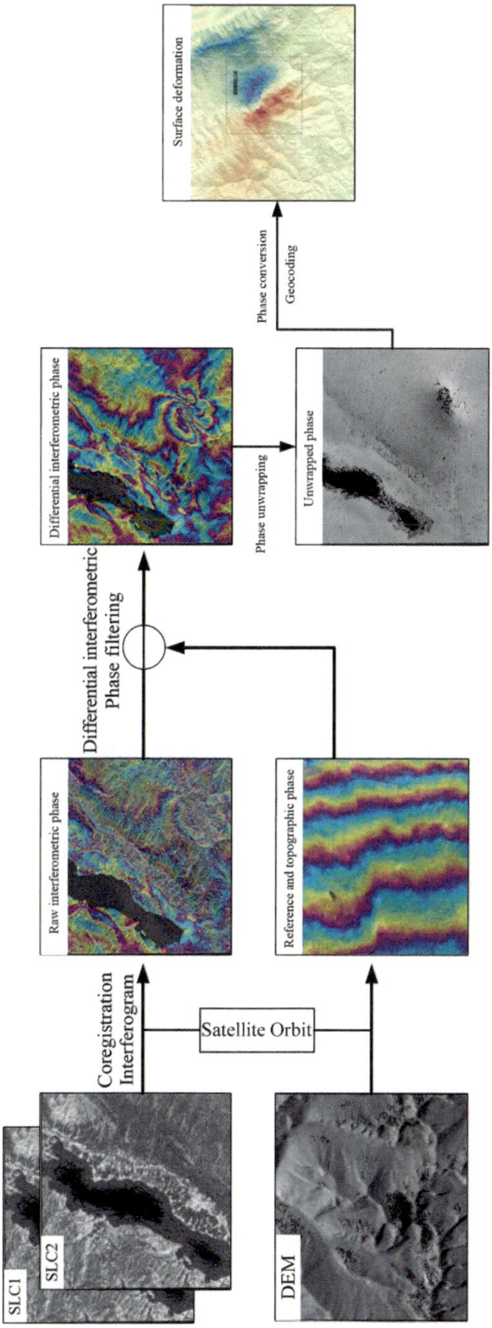

Fig. 2.4 Workflow of D-InSAR processing

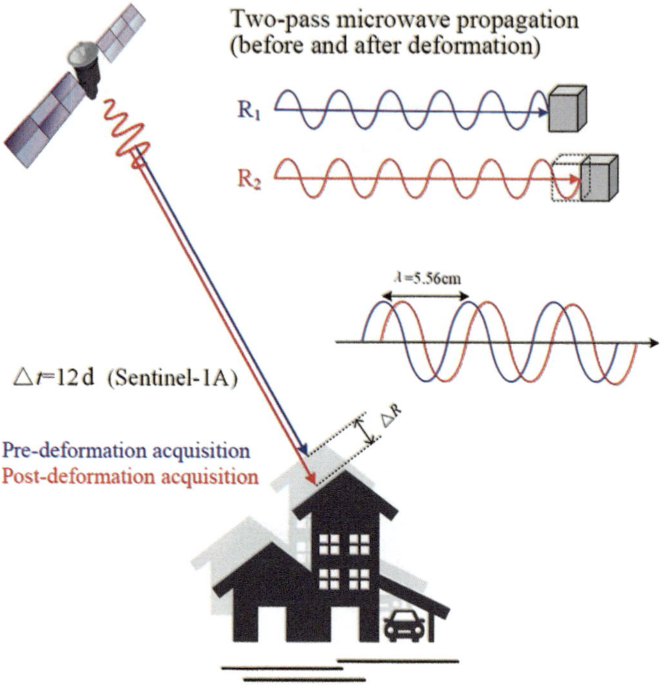

Fig. 2.5 Principles of time-series InSAR

selecting an appropriate time-series of images, temporal information about surface deformation can be derived.

The principles of TS-InSAR involve three key steps: phase difference calculation, deformation measurement computation, and deformation model establishment. Phase difference calculation: using InSAR technology with two or more SAR acquisitions, the phase difference for each SAR data pair is obtained as shown in Eq. (2.14). Deformation measurement conversion: the phase difference is then converted into surface deformation values, representing displacement along the azimuth (along-track) and range (cross-track) directions between each SAR data pair. The formula is:

$$d = (d_{LOS}, d_N, d_E) = \frac{\Delta\phi}{4\pi}\frac{\lambda}{B \cdot n}(B \cdot u) \tag{2.14}$$

Here, d represents the surface deformation vector, with d_{LOS}, d_N, and d_E denoting its components along the LOS direction, northward, and eastward, respectively; n is the normal vector to the Earth's surface; u is the unit vector of the deformation component along the LOS direction; B is the baseline vector between two acquisitions; and λ is the radar wavelength.

Finally, for an area exhibiting time-series deformation, a deformation model can be established based on the cumulative deformation over time. The formula is:

$$d(t) = d_0 + V \vartriangle t + \alpha \frac{\vartriangle t^2}{2} \tag{2.15}$$

Here, $d(t)$ represents the surface deformation vector at time t, d_0 denotes the reference deformation vector at a baseline epoch, V is the deformation velocity vector, and α represents the deformation acceleration vector. By establishing such a deformation model, future surface deformation can be predicted, and potential geohazard-prone areas can be identified. As shown in Fig. 2.6, time-series InSAR technology has been widely applied in monitoring urban land subsidence and infrastructure deformation, including buildings, bridges, highways, railways, subways, factories, and airports (He et al., 2021, 2022, 2023; Yao et al., 2023).

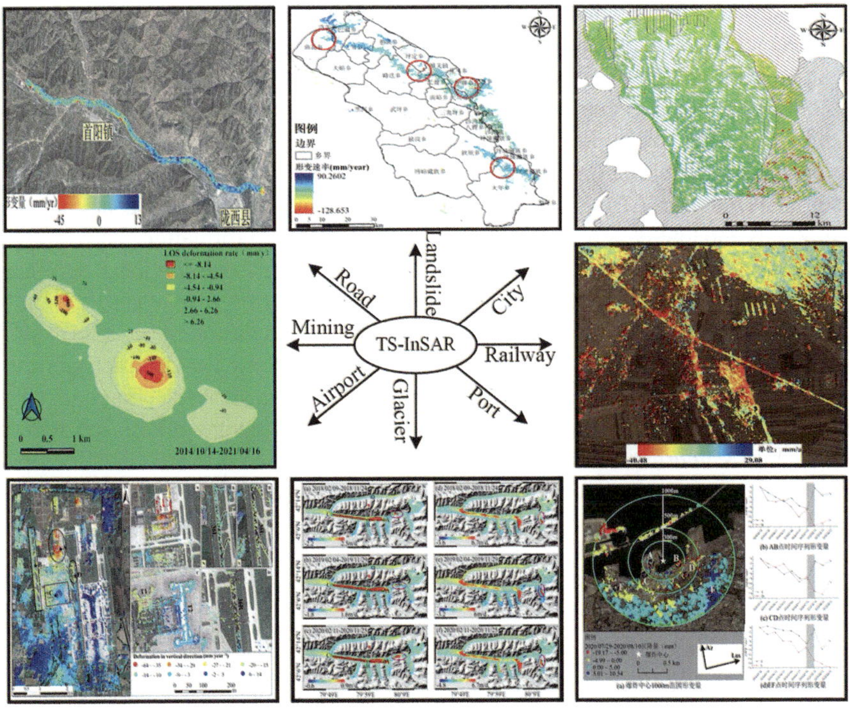

Fig. 2.6 TS-InSAR application cases

2.1.3.1 PS-InSAR

As mentioned above, while traditional D-InSAR technology can effectively detect ground surface deformation and has matured for widespread applications in monitoring earthquakes, volcanoes, and surface deformation, its measurement accuracy remains vulnerable to temporal-spatial decorrelation and atmospheric interference. Spatial decorrelation occurs when excessive spatial baseline length between interferometric pairs leads to significant differences in radar incidence angles, causing substantial changes in target scattering phases. Temporal decorrelation results from noticeable alterations of surface scatterers over short time periods. Atmospheric interference arises from variations in temperature, humidity, and pressure during radar imaging, introducing phase delays and atmospheric phase errors.

To address these limitations, Ferretti et al. (2001) proposed Persistent Scatterer InSAR (PS-InSAR) technology, whose core concept involves identifying high-coherence, stable points called permanent scatterers (PS) based on image amplitude and phase information. These PS points maintain high coherence and strong scattering characteristics over long time series, typically including rocks, man-made structures like buildings and bridges, dams, and corner reflectors. Another key aspect of PS-InSAR involves extracting phases from PS points after differential interferometric processing, constructing neighborhood differential networks, and modeling/solving linear deformation and elevation error parameters. The residual errors are then analyzed considering atmospheric effects as random high-frequency signals temporally and low-pass signals spatially, while nonlinear deformation phases appear as low-frequency signals temporally. Atmospheric phases and nonlinear deformation can be estimated through spatial and temporal filtering, with each PS point's deformation time series obtained by combining linear and nonlinear deformation components. Figure 2.7 shows the basic PS-InSAR workflow. The specific algorithm involves: obtaining N + 1 SAR images covering the same area, selecting one reference image for co-registration, resampling, flat-Earth and topographic phase removal to generate N interferograms. Using reference DEM and satellite orbit data for topographic correction, the interferometric phase at position x in the ith interferogram can be expressed as:

$$\Delta\phi_x(x,y) = \phi(t_B,x,y) - \phi(t_A,x,y) \approx \Delta\phi_{defo,(x,y),i} + \Delta\phi_{topo,(x,y),i} + \Delta\phi_{atm,(x,y),i} + \Delta\phi_{noise,(x,y),i} \tag{2.16}$$

In the equation, the sum of the three components ($\phi_{const,x,i}$, $\phi_{slope,x,i}$, and $\phi_{atm,x,i}$) called the Atmospheric Phase Screen (APS), representing the constant phase value, linear phase component induced by atmospheric effects, and nonlinear atmospheric phase variations, respectively. $\phi_{defo,x,i}$ denotes the surface deformation phase along the radar line-of-sight direction, $\phi_{topo,x,i}$ represents the topographic phase caused by DEM errors, and $\phi_{noise,x,i}$ indicates the noise phase introduced by factors such as scatterer changes, thermal noise, and co-registration errors.

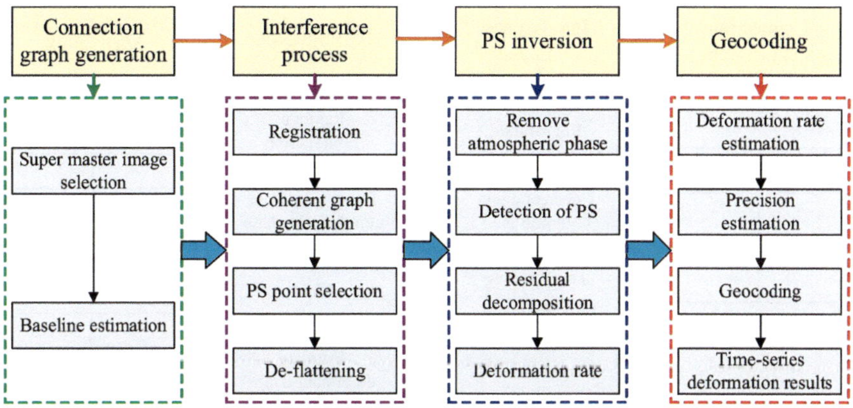

Fig. 2.7 Basic processing flow of PS-InSAR, reprinted from Ref. He et al. (2021), copyright 2025, with permission from Elsevier

To acquire the final ground surface deformation components, a differential interferometric phase model is established for PS points, where each phase component is separated according to its spatiotemporal characteristics, ultimately obtaining both linear and nonlinear deformation variables at the PS points.

2.1.3.2 SBAS-InSAR

Compared to traditional D-InSAR technology, PS-InSAR technology effectively addresses issues such as temporal and spatial decorrelation and atmospheric interference inherent in D-InSAR. It is widely utilized for monitoring surface deformation, with accuracy typically reaching the millimeter level and even achieving sub-millimeter precision in areas exhibiting linear deformation. PS-InSAR identifies permanent scatterers to monitor their deformation; however, in non-urban areas, the scarcity of strong scatterers limits its deformation detection capability, making it nearly impossible to obtain effective monitoring results. Furthermore, PS-InSAR generally employs a linear deformation model to extract corresponding time-series deformation results from target points. In regions exhibiting significant nonlinear deformation characteristics, the deformation process may exceed the boundary conditions of the model. In such cases, utilizing a linear deformation model to inversely estimate the deformation rate can lead to substantial discrepancies between the estimated results and the actual deformation conditions.

To overcome the limitations of PS-InSAR technology, Berardino et al. (2002), proposed the Small Baseline Subset InSAR (SBAS-InSAR) technique. This technique involves grouping all SAR images into small baseline subsets according to predefined spatial and temporal baseline thresholds, thereby ensuring that the baseline distances within each subset are minimized. For each subset, the least squares method is used to estimate the time series of surface deformation. The Singular Value

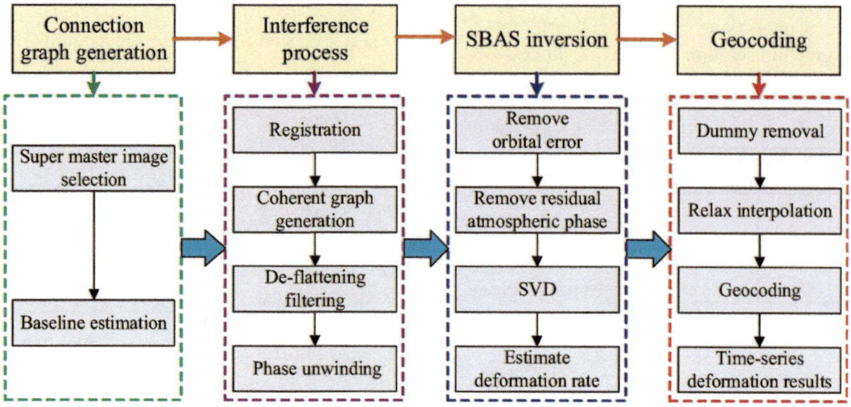

Fig. 2.8 Basic process flow of SBAS-InSAR processing, reprinted from Ref. He et al. (2021), copyright 2025, with permission from Elsevier

Decomposition (SVD) method is employed to address potential ill-posed parameter estimation issues between subsets. By combining multiple interferograms, a linear deformation model is established based on highly coherent points to invert and obtain information such as linear deformation rates and elevation errors. The residual errors are then subjected to spatiotemporal filtering to remove atmospheric delays and are decomposed to extract nonlinear deformation information. Figure 2.8 illustrates the basic process of the SBAS-InSAR technique. The specific algorithm is as follows:

$$\Delta\phi_x(x,y) = \phi(t_B,x,y) - \phi(t_A,x,y) \approx \Delta\phi_{defo,(x,y),i} + \Delta\phi_{topo,(x,y),i} + \Delta\phi_{atm,(x,y),i} + \Delta\phi_{noise,(x,y),i} \tag{2.17}$$

Acquire N + 1 SAR images covering the same region. Select one image as the reference image and perform registration, resampling, flat-earth removal, and terrain phase removal. Based on reasonable spatial and temporal baseline thresholds, M interferograms are generated, which must satisfy the inequality $(N+1)/2 < M < N(N+1)/2$. For the i-th interferogram (i = 1,…,M), the phase value of each pixel is given by.

In the equation, t_A, t_B ($t_A > t_B$)represents the acquisition time of the SAR image corresponding to the i-th interferogram, $\Delta\phi_{defo,(x,y),i}$ denotes the deformation phase corresponding to the slant range from time t_B to t_A, $\Delta\phi_{topo,(x,y),i}$ is the topographic phase error, $\Delta\phi_{atm,(x,y),i}$ is the atmospheric phase error, and $\Delta\phi_{noise,(x,y),i}$ is the noise phase error.

2.2 Basic Principles of Deep Learning

Deep learning is a machine learning method whose fundamental principle is to simulate the neuron structure of the human brain by constructing and training deep neural network models, learn feature representations from data, and adjust network parameters through optimization algorithms to minimize the loss function, thereby enabling the analysis and processing of various machine learning tasks. In recent years, with the advancement of deep learning, its application in the field of remote sensing has become increasingly widespread, including tasks such as remote sensing image classification, object detection and recognition, and remote sensing image fusion.

Classical algorithms in the field of traditional machine learning, such as support vector machines (SVM), logistic regression (LR), and decision trees, all fall within the category of shallow structure models. These models have certain limitations: when the scale of data samples is limited, their ability to express complex functional relationships is insufficient, and their generalization performance is easily constrained in complex problem scenarios (Bengio, 2009). For highly nonlinear functional mapping relationships, traditional machine learning techniques often struggle to model them effectively. Compared to traditional machine learning techniques, deep learning, which relies on deep neural network architectures, can effectively overcome the aforementioned limitations. Deep neural networks have significantly enhanced the fitting capability of complex functions by introducing richer types of network structures and deeper network layers. In recent years, with the rapid evolution of deep learning technologies, the number of layers in deep neural networks has escalated to tens or even hundreds of layers.

In the field of machine learning, "features" are abstract data sets that reflect the characteristics of raw data, a concept that is of paramount importance. Traditional machine learning methods rely on domain experts to manually design features based on experience, and the quality of these features directly impacts the performance of the model, introducing significant uncertainty due to human intervention. In contrast, deep learning automates feature construction through deep neural networks. This technical approach abandons the traditional mode of manually designing features, instead extracting features directly from raw data through the model, known as representation learning or feature learning. When the scale of training data is sufficient, the high-level features automatically generated by deep neural networks often surpass the effectiveness of manually designed features. The end-to-end architecture of deep learning (outputting results from raw data) possesses greater potential for global optimization compared to the traditional multi-step divide-and-conquer architecture (e.g., feature design, extraction, selection, and result output). This characteristic enables it to demonstrate broader development prospects in multiple application fields.

2.2.1 Neural Networks

A neural network is a system model that mimics biological neurons, composed of densely interconnected simple nonlinear analog processing elements known as nodes. The network is formed by connecting the output of each node to the inputs of some other nodes, similar to the synaptic connections of real neurons. Each neuron represents a specific output function, called an activation function, and each connection between two neurons includes a connection strength, which is the weighted value acting on the signal passing through that connection. After training, a neural network possesses the capabilities of information feature extraction, knowledge generalization, and learning and memory, with the information or knowledge learned by the model stored in the connection matrix between each unit node. The performance of a neural network is collectively determined by three characteristics: node definition, topology, and learning algorithm.

A neuron is generally a nonlinear structure with multiple inputs and a single output. As the basic computational unit of a neural network, it can have feedback inputs and threshold parameters. A general unit model is shown in Fig. 2.9.

Every input connection X_j has a connection strength, which is amplified or attenuated by the connection weight W_{ji}, known as weighting. The neuron processes the weighted input signals quantitatively, then sums them with value b_i to calculate the total weighted input value, which determines a unique output V_i. This output is then passed through a function $f(\cdot)$ to obtain the final output value Y_i.

$$V_i = \sum_{j=1}^{N} X_j W_{ji} + b_i \tag{2.18}$$

$$Y_i = f(V_i) \tag{2.19}$$

Herein, $V(\cdot)$ represents the basis function, which is a single-output, multiple-input function. $V = V(X, W, b), f(\cdot)$ is referred to as the activation function, also known as the transfer function or transmission function. The general role of the activation function $f(\cdot)$ is to squeeze the output of the basis function $V(\cdot)$, as shown in Eq. (2.20), where the nonlinear function $f(\cdot)$ transforms V_i into a specified range.

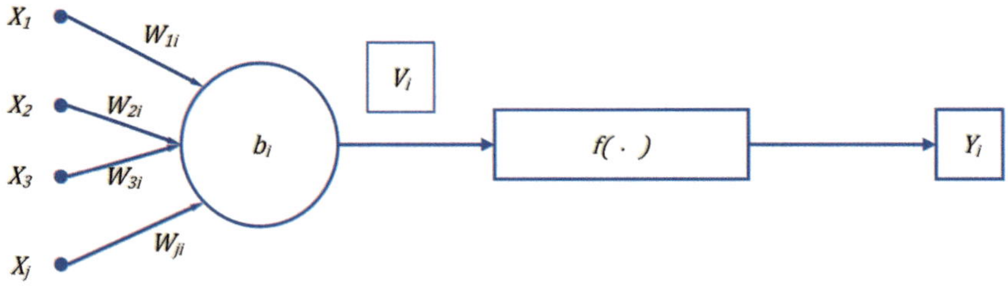

Fig. 2.9 General neuron model

The commonly used basis functions include:

(1) Linear function

$$y = f(u) = u \tag{2.20}$$

(2) Distance function. The output of this basis function is:

$$V_i = \sqrt{\sum_{j=1}^{N} (X_j - W_{ji})^2} = ||\boldsymbol{X} - \boldsymbol{W}|| \tag{2.21}$$

In the formula, $\boldsymbol{W}$ is often referred to as the center of the basis function. Clearly, V_i represents the Euclidean distance between the input vector $\boldsymbol{X}$ and the weight vector $\boldsymbol{W}$. In a multidimensional space, the shape of this basis function is a hypersphere with $\boldsymbol{W}$ as its center, primarily used in Radial Basis Function (RBF) neural networks.

(3) Elliptic basis function. The output of this function is:

$$V_i = \sqrt{\sum_{j=1}^{N} c_i (X_j - W_{ji})^2} \tag{2.22}$$

In a multidimensional space, the shape of this basis function is an ellipsoid.
There are various choices for the activation function, commonly used ones include:

(1) Hard limit function. The function is as follows:

$$y = f(x) = \begin{cases} 0, x < 0 \\ 1, x \geq 0 \end{cases} \tag{2.23}$$

or

$$y = f(x) = sgn(x) = \begin{cases} -10, x < 0 \\ 1, x \geq 0 \end{cases} \tag{2.24}$$

In the formula, $sgn(\cdot)$ represents the function symbol. The first one is a single-limit function, and the second one is a double-limit function.

(2) Linear function. The function expression is as follows:

$$y = f(x) = x \tag{2.25}$$

This activation function is commonly used in the output neurons of neural networks that implement function approximation.

(3) The Sigmoid function, also known as the S-shaped function, is strictly monotonically increasing. The logarithmic tangent function is a form of the S-shaped

function. The function expression is as follows:

$$y = f(x) = \frac{1}{1 + e^{-\lambda x}} \tag{2.26}$$

In the formula, parameter λ represents the gain of the Sigmoid function, which is differentiable and varies continuously within the range of 0 to 1.

2.2.2 *Convolutional Neural Network*

Lecun et al. (1998) first proposed the convolutional neural network (CNN), which has now been widely applied in the field of computer vision and has achieved numerous results, making it one of the representative algorithms of deep learning. The basic network structure of CNN mainly includes the input layer, hidden layer, and output layer, with the hidden layer consisting of convolutional layers, pooling layers, and fully connected layers (Fig. 2.10). CNN can learn task-appropriate feature representations from a large amount of image data without the need for manual feature design. Its application in the field of image processing reduces the number of weights, making the network easier to optimize, and lowers the complexity of the model, thereby reducing the risk of overfitting.

(1) Convolutional layer

The convolutional layer primarily extracts target features by performing convolution operations between the convolutional kernel and the input samples, and it is the core component of CNN. The convolutional kernels used in convolution operations are different, and the parameters on each convolutional kernel are learned and updated through a feedback mechanism. The convolutional kernel moves regularly across the image, and the corresponding regions complete the convolution operation, thereby obtaining convolutional features. These convolutional features typically possess the characteristics of local perception and weight sharing.

The convolution operation can be expressed as:

$$C_i = f(W_i \otimes C_{i-1} + b_i) \tag{2.27}$$

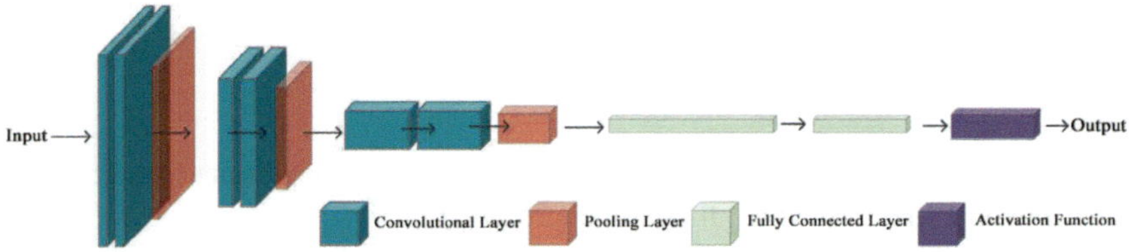

Fig. 2.10 Basic structure of CNN

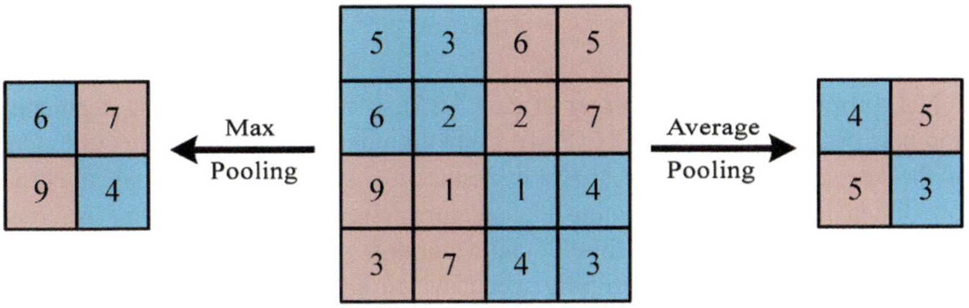

Fig. 2.11 Pooling operation

In the formula, W_i represents the weight of the convolution kernel, $\otimes$ represents the convolution operation, b_i is the bias term, and $f(\bullet)$ is the activation function operation.

(2) Pooling layer
The pooling layer primarily performs down-sampling on the feature maps obtained after the convolution operation, reducing the size of the feature maps to decrease the number of network parameters and computational load. The pooling layer can simplify the representation of semantic information, increase the receptive field, and enhance robustness to noise processing. Common pooling operations include average pooling and max pooling, and the operation process is illustrated in Fig. 2.11.

(3) Fully connected layer
The primary function of the fully connected layer is to integrate the features extracted by the convolutional and pooling layers, and to perform inference and prediction to obtain the final result. The fully connected layer enhances the model's expressive and fitting capabilities by multiplying the input feature vector with the weight matrix, adding the bias term, and then applying an activation function for nonlinear transformation, thereby mapping the result to the final output value. Common activation functions include Sigmoid and ReLU.

The calculation formula for Sigmoid is:

$$Sigmoid(x) = \frac{1}{1 + e^{-x}} \tag{2.28}$$

The calculation formula for ReLU is:

$$SReLU(x) = max(0, x) \tag{2.29}$$

2.2.3 U-Net Network

U-Net, proposed by Ronneberger et al. (2015), is an image semantic segmentation network based on the FCN (Fully Convolutional Networks) architecture. One of the key ideas of U-Net is the skip connection structure, which can integrate low-dimensional and high-dimensional features, significantly improving segmentation accuracy. At the same time, it can accurately segment images with minimal training data and has a fast training speed, which has led to its widespread application in remote sensing image segmentation. The architecture of U-Net consists of a contracting path (encoder), an expanding path (decoder), and skip connection structures, forming a "U"-shaped structure, which is a typical encoder-decoder architecture (Lu et al., 2023).

The encoder is used to extract features from images and reduce spatial dimensions. Similar to a standard CNN architecture, the encoder comprises convolutional layers and downsampling layers. Convolutional layers are employed to extract image features, while downsampling layers are utilized to filter out insignificant high-frequency information and perform feature dimensionality reduction. Repeated convolutional and pooling operations can thoroughly extract high-level features of images. The role of the decoder is to upsample the feature maps obtained by the encoder, aligning them with the dimensions of the input image, thereby enabling pixel-level semantic prediction of the input image. The decoder includes convolutional layers, upsampling layers, and concatenation operations. During the decoding process, each time upsampling is performed, the number of channels in the feature map is halved, and then the feature maps of the corresponding scales from the encoding part and the decoding part are fused through a dimensional concatenation operation. Dimensional concatenation involves concatenating the channel dimensions of the feature maps together to form a thicker feature map. This operation can preserve more high-resolution detail information, thereby improving the resolution and edge accuracy of the final segmentation result. In this paper, U-Net is adopted as the base network for the landslide identification and extraction model. In this paper, the number of channels in the adopted U-Net starts from 32 and becomes 512 after 4 downsampling operations. Its structure is shown in Fig. 2.12.

The skip connection structure of U-Net combines shallow, low-level mapping from the encoder with deep, high-level features from the decoder, thereby enhancing segmentation accuracy. However, research indicates that when the segmentation target is very small, the segmentation performance is suboptimal, as the skip connection structure directly merges shallow features with deep features, which can easily lead to a semantic gap. Although the skip connection structure is a natural design, the feature maps of the same scale in the encoder and decoder networks are semantically different, and it cannot be guaranteed that they are the best match for feature fusion (Abdollahi & Pradhan, 2021).

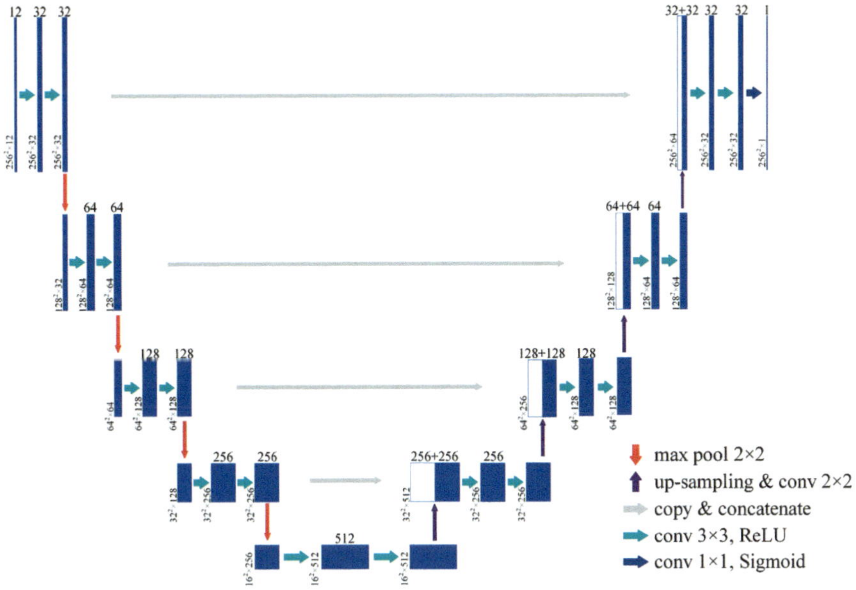

Fig. 2.12 U-Net network architecture

2.2.4 *Gated Recurrent Unit*

GRU is a type of Recurrent Neural Network (RNN) that addresses issues in RNNs such as the inability to retain long-term memory and problems with gradients during backpropagation. The GRU architecture simplifies LSTM by merging the forget gate and input gate into a single update gate, using a single gating unit to simultaneously determine the forgetting factor and update the state. The network consists of only two gate functions: the update gate and the reset gate. The update gate allows for conditional leakage accumulation, enabling linear control over any dimension to either replicate or update it; the reset gate controls the portion of the current state used for the next state, introducing additional nonlinear effects between sequences. The specific functions are as follows: r represents the reset gate, z represents the update gate, $h^{(t-1)}$ denotes the output state at the previous time step, $\tilde{h}^{(t)}$ represents the hidden state information at the current time step, and $h^{(t)}$ denotes the output value at the current time step. Its structure is shown in Fig. 2.13.

$$r^{(t)} = \sigma\left(W_r x^{(t)} + U_r h^{(t-1)} + b_r\right) \tag{2.30}$$

$$z^{(t)} = \sigma\left(W_z x^{(t)} + U_z h^{(t-1)} + b_z\right) \tag{2.31}$$

$$\tilde{h}^{(t)} = \tan h\left(W_h x^{(t)} + U_h\left(r^{(t)} \circ h^{(t-1)}\right)\right) \tag{2.32}$$

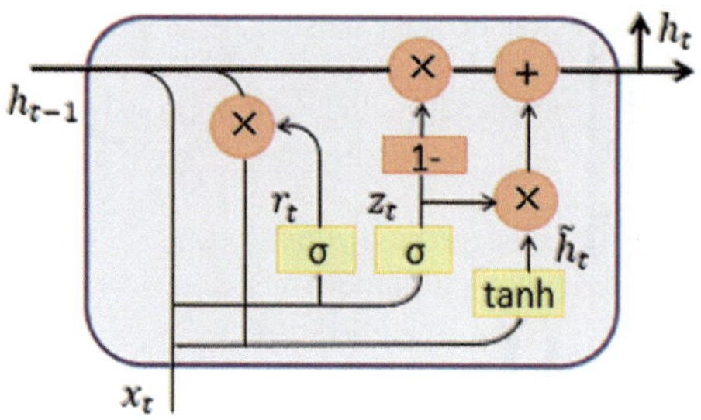

Fig. 2.13 GRU neuron structure

$$h^{(t)} = z^{(t)} \circ h^{(t)} + \left(1 - z^{(t)}\right) \circ h^{(t-1)} \quad (2.33)$$

Here, $\sigma(\cdot)$ represents the sigmoid function, $\tan h(\cdot)$ is the hyperbolic tangent function, $x^{(t)}$ is the input at time t; $\circ$ denotes the Hadamard product; W and U represent the weight matrices, and b denotes the bias.

2.3 Summary

This chapter focused on the fundamental principles of Synthetic Aperture Radar Interferometry (InSAR) and deep learning techniques. It introduces the basic principles of D-InSAR deformation monitoring technology and neural networks. The chapter provides a detailed explanation of the fundamental principles of time-series InSAR deformation monitoring technology, including two classic time-series InSAR deformation monitoring techniques: SBAS-InSAR and PS-InSAR. It also elaborates on three commonly used neural networks: Convolutional Neural Networks (CNN), U-Net networks, and GRU networks.

References

Abdollahi, A., & Pradhan, B. (2021). Integrating semantic edges and segmentation information for building extraction from aerial images using UNet. *Machine Learning with Applications, 6*, Article 100194. https://doi.org/10.1016/j.mlwa.2021.100194

Baran, I., Stewart, M., & Claessens, S. (2005). A new functional model for determining minimum and maximum detectable deformation gradient resolved by satellite radar interferometry. *IEEE Transactions on Geoscience and Remote Sensing, 43*(4), 675–682. https://doi.org/10.1109/TGRS.2004.843187

Bengio, Y. (2009). Learning deep architectures for AI. *Foundations and Trends® in Machine Learning*, 2(1), 1–127. https://doi.org/10.1561/2200000006

Berardino, P., Fornaro, G., Lanari, R., & Sansosti, E. (2002). A new algorithm for surface deformation monitoring based on small baseline differential SAR interferograms. *IEEE Transactions on Geoscience and Remote Sensing, 40*(11), 2375–2383. https://doi.org/10.1109/TGRS.2002.803792

Brasca Merlin, A., Solarte, A., Bellis, L. M., Carignano, C., Cioccale, M., Delgado, M., & Argañaraz, J. P. (2021). DInSAR and statistical modeling to assess landslides: The case study of Sierras Chicas (central Argentina). *Journal of South American Earth Sciences, 108*, Article 103179. https://doi.org/10.1016/j.jsames.2021.103179

Cai, J, Zhang, L, Dong, J, et al. (2020). Detection and monitoring of post-earthquake landslides in Jiuzhaigou using radar remote sensing. *Geomatics and Information Science Of Wuhan University, 45*(11), 1707–1716.

Chen, Q., Yang, Y., Liu, G., Cheng, H., & Liu, W. (2012). InSAR least squares integer phase unwrapping method based on boundary detection. *Acta Geodaetica Et Cartographica Sinica, 41*(4), 441–448. [In Chinese].

Ferretti, A., Prati, C., & Rocca, F. (2001). Permanent scatterers in SAR interferometry. *IEEE Transactions on Geoscience and Remote Sensing, 39*(1), 8–20. https://doi.org/10.1109/36.898661

Feynman, R. P. (1963). The Feynman lectures on physics (Vol. 1). Addison-Wesley.

Hanssen, R. F. (2001). Radar interferometry: Data interpretation and error analysis. Springer.

He, Y., Wang, W., Zhang, L., Chen, Y., Chen, Y., Chen, B., Zhao, Z. (2023). An identification method of potential landslide zones using InSAR data and landslide susceptibility. *Geomatics, Natural Hazards and Risk*, 14(1). https://doi.org/10.1080/19475705.2023.2185120

He, Y., Chen, Y., Wang, W., Yan, H., Zhang, L., & Liu, T. (2021). TS-InSAR analysis for monitoring ground deformation in Lanzhou new district, the loess Plateau of China, from 2017 to 2019. *Advances in Space Research, 67*(4), 1267–1283. https://doi.org/10.1016/j.asr.2020.11.004

He, Y., Yan, H., Yang, W., Yao, S., Zhang, L., Chen, Y., & Liu, T. (2022). Time-series analysis and prediction of surface deformation in the Jinchuan mining area, Gansu province, by using InSAR and CNN-PhLSTM network. *IEEE Journal of Selected Topics in Applied Earth Observations and Remote Sensing, 15*, 6732–6751. https://doi.org/10.1109/JSTARS.2022.3198728

Howard, H. R., Manandhar, S., Wang, Q., McMillan, J. M., Qie, G., Liu, X., & Wang, G. (2022). Spatially characterizing land surface deformation and permafrost active layer thickness for Donnelly installation of Alaska using DInSAR and MODIS data. *Cold Regions Science and Technology, 196*, Article 103510. https://doi.org/10.1016/j.coldregions.2022.103510

Jordan, R. L. (1980). The Seasat-A synthetic aperture radar system. *IEEE Journal of Oceanic Engineering, 5*(2), 154–164. https://doi.org/10.1109/JOE.1980.1145458

Kothyari, G. C., Mishra, S., Taloor, A. K., Kandregula, R. S., Pathak, V., & Chauhan, G. (2022). Distribution of neotectonic variability between the Kachchh Mainland Fault and Vigodi Fault, Northwestern Mainland Kachchh, Western India. *Quaternary Science Advances, 6*, Article 100047. https://doi.org/10.1016/j.qsa.2021.100047

Kumar, A., Manisha, Lal, P., Prasad, A., Tripathy, P., Saikia, P. (2022). Analyzing urban damage and surface deformation based hazard-risk in Kathmandu city occurred during Nepal earthquake (2015) using SAR interferometry. *Advances in Space Research*, 70(12), 3892-3904. https://doi.org/10.1016/j.asr.2022.02.003

Lecun, Y., Bottou, L., Bengio, Y., & Haffner, P. (1998). Gradient-based learning applied to document recognition. *Proceedings of the IEEE, 86*(11), 2278–2324. https://doi.org/10.1109/5.726791

Liu, G. (2019). *Principles and Applications of InSAR* (In Chinese). Science Press.

Liu, J., Hu, J., Li, Z. et al. (2021). Estimation of 3D coseismic deformation with InSAR: an improved SM-VCE method by window optimization. *Acta Geodaetica et Cartographica Sinica, 50*(9), 1222–1239.

Lu, W., Hu, Y., Zhang, Z., & Cao, W. (2023). A dual-encoder U-Net for landslide detection using Sentinel-2 and DEM data. *Landslides, 20*(9), 1975–1987. https://doi.org/10.1007/s10346-023-02089-5

Niraj, K. C., Gupta, S. K., & Shukla, D. P. (2022). Kotrupi landslide deformation study in non-urban area using DInSAR and MTInSAR techniques on Sentinel-1 SAR data. *Advances in Space Research, 70*(12), 3878–3891. https://doi.org/10.1016/j.asr.2021.11.042

Ronneberger, O., Fischer, P., & Brox, T. (2015). U-Net: Convolutional networks for biomedical image segmentation. In N. Navab, J. Hornegger, W. Wells, & A. Frangi (Eds.), *Medical image computing and computer-assisted intervention—MICCAI 2015* (pp. 234–241). Springer. https://doi.org/10.1007/978-3-319-24574-4_28

Vélez, M., Bustos, E., Euillades, L., et al. (2021). Ground deformation at the Cerro Blanco caldera: A case of subsidence at the Central Andes BackArc. *Journal of South American Earth Sciences, 106*, 102941.

Wright, T. J., Parsons, B., England, P. C., & Fielding, E. J. (2004). InSAR observations of low slip rates on the major faults of western Tibet. *Science, 305*(5681), 236–239. https://doi.org/10.1126/science.1096388

Yao, S., He, Y., Zhang, L., Yang, W., Chen, Y., Sun, Q., & Cao, S. (2023). A ConvLSTM neural network model for spatiotemporal prediction of mining area surface deformation based on SBAS-InSAR monitoring data. *IEEE Transactions on Geoscience and Remote Sensing, 61*, 1–22. https://doi.org/10.1109/TGRS.2023.3236510

Yao, X., Chen, Y., Liu, D., Zhou, Z., Liesenberg, V., Marcato Junior, J., & Li, J. (2021). Average-DInSAR method for unstable escarpments detection induced by underground coal mining. *International Journal of Applied Earth Observation and Geoinformation, 103*, Article 102489. https://doi.org/10.1016/j.jag.2021.102489

Chapter 3
Deep Learning Approaches for Intelligent Landslide Detection

This chapter systematically introduces intelligent landslide recognition methods, including the attention mechanism U-Net landslide recognition method based on optical remote sensing, the multi-input channel U-Net landslide recognition method based on the fusion of multi-source SAR remote sensing data, and the multi-model ensemble landslide recognition method that considers spectral, topographic, and textural features. The typical study areas are Lanzhou and the Bailong river basin, where landslide disasters occur frequently in Northwest China.

3.1 Attention Mechanism U-Net for Landslide Detection Using Optical Remote Sensing

To address the limitations posed by insufficient information in remote sensing imagery and the U-Net model's limited ability to emphasize the relative importance of feature channels—both of which result in poorly defined landslide features—this chapter introduces multi-band Sentinel-2A optical imagery for landslide detection. A channel attention mechanism-based U-Net model is developed to enhance the generalization and stability of landslide recognition, thereby enabling intelligent and accurate detection.

3.1.1 Experimental Setup and Data Sources

3.1.1.1 Study Area

The geological setting of Lanzhou is highly complex. As illustrated in Fig. 3.1, much of the region consists of loess-covered hills and basins. Located at 35.57°N

Y. He, *InSAR and Deep Learning in Landslides Research: Intelligent Identification, Risk Assessment and Susceptibility Mapping*, https://doi.org/10.1007/978-981-96-9132-6_3

and 102.59°E, Lanzhou lies within the transitional zone between the Qinghai–Tibet Plateau and the Loess Plateau, characterized by significant topographic variability and intricate geological conditions. The city extends along the east–west course of the Yellow River, with mountains flanking both the north and south, exhibiting the morphology of a belt-shaped basin city. Lanzhou experiences a temperate continental climate, with hot, rainy summers and cold, dry winters. The urban area is predominantly covered by Quaternary aeolian loess, which is vertically jointed and loosely structured, making it highly susceptible to landslide during the rainy season. Furthermore, intensified anthropogenic activities in recent years have exacerbated the impact on the geological environment, resulting in frequent landslide occurrences (Fang et al., 2011). The dominant landslide types in Lanzhou are hybrid, mid-layer, and small-scale loess landslides. These often occur in loess or loess-like soils, where they are difficult to distinguish from surrounding surface features, and the landslide boundaries are often ill-defined (Mei & Zhang, 2010). Additionally, as a provincial capital, Lanzhou is characterized by a dense transportation network and high population mobility. Intensive activities such as infrastructure development and mineral extraction have further contributed to landslide (Wang et al., 2014). The occurrence of landslides poses substantial threats to both residents and urban infrastructure. Consequently, early identification and management of landslide-prone zones in such disaster-susceptible cities—represented here by the core urban area of Lanzhou—are urgently needed. This study focuses on the central urban area of Lanzhou, comprising Chengguan, Anning, Xigu, and Qilihe districts, for landslide detection and susceptibility mapping.

3.1.1.2 Data and Preprocessing Methods

This study employed Sentinel-2A satellite imagery acquired from the official platform of the European Space Agency (ESA: https://scihub.copernicus.eu/, accessed on May 6, 2025). The Sentinel-2A sensor provides 13 spectral bands spanning the visible to near-infrared range, with spatial resolutions of 10 m, 20 m, and 60 m, offering rich spectral information. To mitigate the influence of cloud contamination on image quality, a strict cloud coverage threshold of 0% was applied during data selection. Accordingly, a cloud-free Sentinel-2A image of Lanzhou acquired on December 5, 2021, was selected for analysis.

The downloaded data were Level-1C (L1C) products, which had not been atmospherically corrected. To generate surface reflectance products (Level-2A), the Sen2cor processor—an ESA-recommended tool—was employed to perform atmospheric correction. The resulting L2A data were then resampled to a uniform 10 m spatial resolution using SNAP software. Twelve bands, excluding Band 10, were retained and exported in a format compatible with ENVI software for further processing.

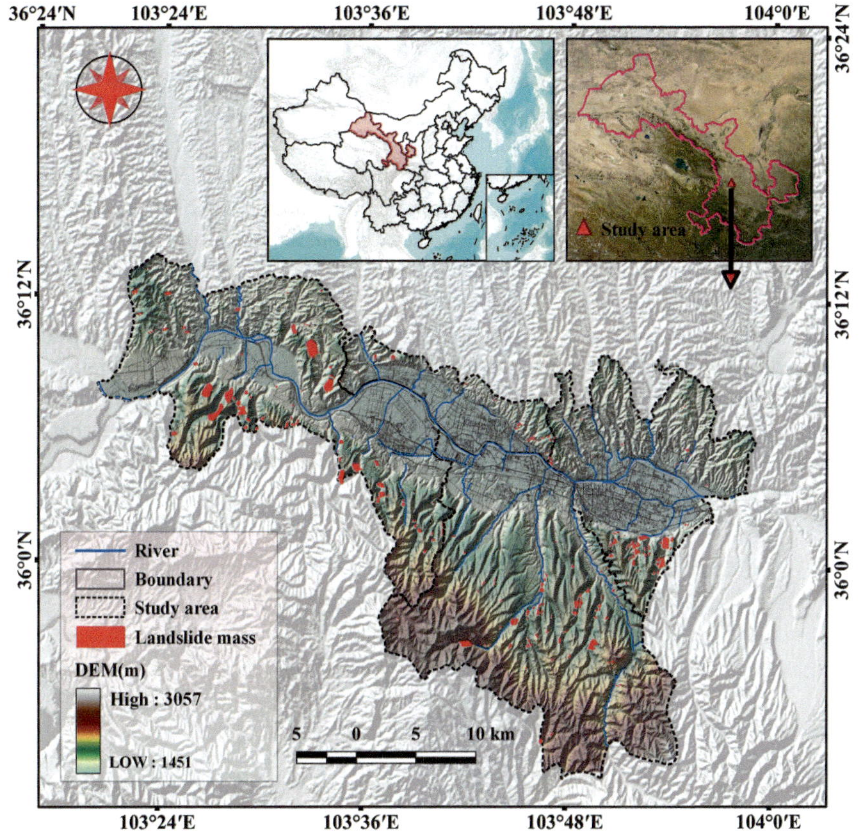

Fig. 3.1 Study area (Lanzhou, Northwest China), reprinted from Ref. Chen et al. (2023), copyright 2025, with permission from Taylor & Francis

3.1.1.3 Dataset Creation and Partitioning

For model training, historical landslide data were used to manually annotate landslide boundaries within the study area through visual interpretation in Google Earth, resulting in 255 delineated landslide polygons. These vector annotations were imported into ArcMap 10.7 and converted into raster format, which were then exported as binary masks. In these masks, white pixels (foreground) represent landslide-affected areas, while black pixels (background) denote non-landslide areas, serving as ground truth labels for supervised deep learning.

The 12 spectral bands of the preprocessed Sentinel-2A imagery were imported into ArcMap 10.7 and spatially aligned with the experimental area, resulting in twelve study-area-specific images. Due to the large size of the original satellite data, which could exceed the memory capacity of standard computing systems, a Python-based cropping script was employed. A sliding window approach with a step size of 128 pixels was used to segment both the original Sentinel-2A images and the corresponding ground truth landslide samples into smaller patches of 256 × 256 pixels, followed by normalization. Each patch contained one or more landslide events. In deep learning, sufficient training samples are essential for model performance. Given the limited availability of high-quality and representative training data in this study, several data augmentation strategies were applied to expand the dataset and improve model generalizability and robustness. Specifically, image patches were augmented through mirroring and 90° counterclockwise rotation, and the same operations were performed on the corresponding landslide masks. This process yielded a total of 5,448 training samples, each with dimensions of 256 × 256 pixels.

The dataset was randomly partitioned into training, validation, and test subsets at a ratio of 6:2:2. This resulted in 3,270 samples for training, 1,089 for validation, and 1,089 for testing. The three subsets were mutually exclusive (Fig. 3.2). The training set was used to fit the model parameters, the validation set to tune hyperparameters and prevent overfitting, and the test set to quantitatively assess the final model performance.

3.1.2 Construction of the Channel Attention Mechanism U-Net for Landslide Detection

3.1.2.1 Fundamental Principles of SENet

In this study, a channel attention mechanism, SENet (Squeeze-and-Excitation Networks) (Hu et al., 2018), was integrated after each skip connection in the U-Net architecture. This mechanism estimates the contribution of each feature channel to landslide classification and enhances or suppresses them accordingly, thereby improving the model's feature learning capability. The core of SENet is the Squeeze-and-Excitation block (SE-block), as shown in Fig. 3.3.

SENet is designed to capture channel-wise interdependencies and automatically learn channel importance through a network-based approach. It assigns dynamic weights to emphasize informative channels while suppressing less relevant ones, thus enhancing the model's representation capacity. The SENet process can be conceptualized in three main stages:

Squeeze Operation

The first stage of the SE-block, known as the Squeeze operation, compresses spatial information by applying global average pooling to each input feature map U, reducing each two-dimensional feature map to a single scalar value $z \in \mathrm{R}^C$ (Eq. (3.1)). This global descriptor captures the overall response of each channel and retains the same dimensionality as the number of channels in the input.

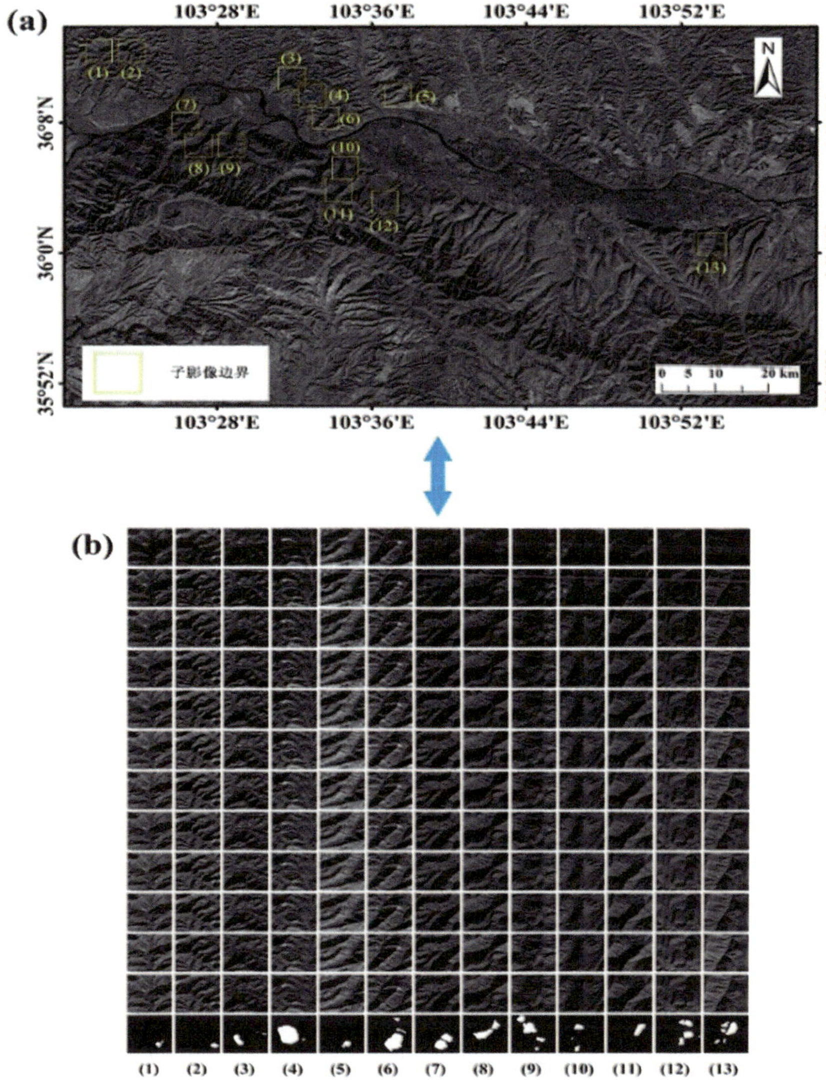

Fig. 3.2 Dataset samples, reprinted from Ref. Chen et al. (2023), copyright 2025, with permission from Taylor & Francis

$$z_c = F_{sq}(u_c) = \frac{1}{H \times W} \sum_{i=1}^{H} \sum_{j=1}^{W} u_c(i, j) \tag{3.1}$$

In this context, u_c denotes a single channel feature map, H and W are the spatial dimensions (height and width), C is the number of channels, and z_c is a scalar value reflecting the global feature of the feature map.

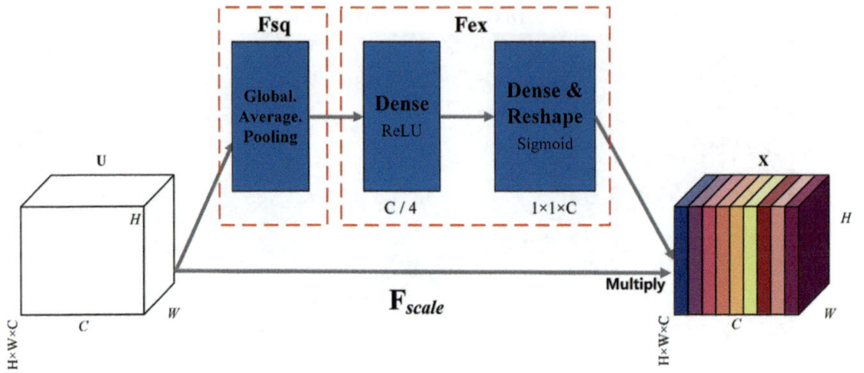

Fig. 3.3 Structure of the SE-block, reprinted from Ref. Chen et al. (2023), copyright 2025, with permission from Taylor & Francis

Excitation Operation

The excitation stage captures inter-channel dependencies and learns nonlinear relationships among feature channels. Functionally analogous to gating mechanisms in recurrent neural networks, this module generates attention weights via a bottleneck architecture composed of two fully connected layers with ReLU and sigmoid activations, respectively (Eq. (3.2)).

$$s = F_{ex}(z, W) = \sigma(g(z, W)) = \sigma(W_2\delta(W_1 z)) \tag{3.2}$$

The transformation parameters are defined as $W_1 \in R^{\frac{c}{r}\times C}$ and $W_2 \in R^{\frac{c}{r}\times C}$, where δ and σ denote the ReLU and sigmoid activation functions. This structure not only reduces model complexity but also improves generalization by learning compact and effective feature representations.

Reweight Operation

In the reweight stage, the channel-wise weights derived from the excitation step are applied to the original feature maps via element-wise multiplication, recalibrating the feature responses along the channel dimension (Eq. (3.3)). This enables the network to emphasize informative features while suppressing less relevant ones, thereby enhancing the model's discriminative capability.

$$\widetilde{\boldsymbol{x}}_c = F_{scale}(\boldsymbol{u}_c, s_c) = s_c \cdot \boldsymbol{u}_c \tag{3.3}$$

In this formula, $\tilde{X} = [\tilde{x}_1, \tilde{x}_2, \ldots, \tilde{x}_C]$, where $\tilde{x}_c$ and $F_{scale}(u_c, s_c)$ represent the product of the scalar s_c and the feature map $u_c \in R^{H\times W}$ on a per-channel basis.

3.1.2.2 Model Construction Approach

The attention mechanism (AM), inspired by human visual processing, mimics the way human vision rapidly scans a scene to locate key regions—attention foci—and then allocates more cognitive resources to those regions to extract fine-grained details while suppressing irrelevant background information. Since its first application in machine translation in 2014, attention mechanisms have been widely adopted in computer vision tasks (Niu et al., 2021). In this domain, attention is typically categorized into spatial, channel-wise, and hybrid attention. Fundamentally, attention mechanisms enable neural networks to prioritize relevant information, typically via learnable weights, thereby enhancing performance by allocating limited computational resources to the most informative parts of the input (Yang, 2020). In the context of this study, attention mechanisms are used exclusively for landslide feature extraction. A major challenge in landslide identification lies in the dominance of background classes—such as vegetation, bare soil, and dry land—that often occupy the majority of an image and are considered non-informative. Moreover, these background elements may share spectral similarities with landslides, further complicating accurate classification.

U-Net's skip connections, which fuse shallow features from the encoder with deep semantic features from the decoder, have been shown to improve segmentation accuracy. However, recent studies suggest that such fusion may be suboptimal when the target object is small, due to a semantic gap between low-level and high-level features (Chen et al., 2023). Although skip connections are widely used, the features at matching spatial resolutions from the encoder and decoder may differ semantically, which can degrade the quality of feature fusion. The concatenation of feature maps along the channel dimension also increases the channel count, potentially introducing redundant information and semantic inconsistency. To address this, we propose integrating a channel attention mechanism, specifically the Squeeze-and-Excitation (SE) module, after each skip connection in U-Net. This module adaptively estimates the importance of each feature channel in landslide classification, thereby enhancing the model's feature representation capabilities. The architecture of the proposed attention-enhanced U-Net is illustrated in Fig. 3.4.

3.1.2.3 Model Parameter Configuration and Training Process

Training deep neural networks requires substantial computational resources (Singaravel et al., 2018). In U-Net, the concatenation of feature maps in the channel dimension leads to increased memory usage. Additionally, intermediate variables and the optimizer's internal parameters contribute to significant GPU memory consumption. The experimental hardware environment included an Intel Core i7-10700F processor, 64 GB of RAM, and an NVIDIA RTX 3090 GPU with 24 GB of memory. The software stack consisted of CUDA version 11.1.105, Python 3.9.8, and TensorFlow 2.7.1.

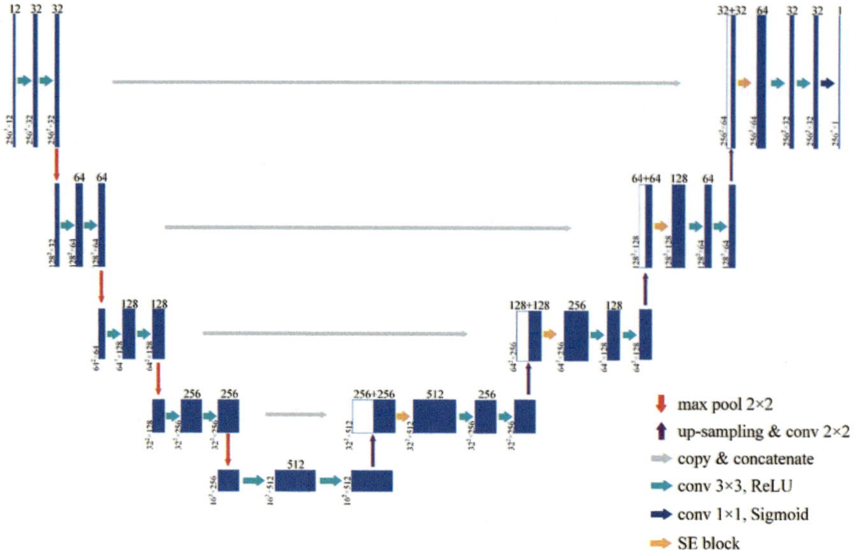

Fig. 3.4 Structure of the Channel Attention Mechanism U-Net Network, reprinted from Ref. Chen et al. (2023), copyright 2025, with permission from Taylor & Francis

Several hyperparameters were empirically optimized. Due to GPU memory constraints, the batch size was set to 64. After initial experimentation, the number of training epochs was set to 128, the learning rate to 0.00001, and the optimizer was chosen as Adam, a widely used adaptive method suitable for non-convex optimization problems in deep learning (Kingma & Ba, 2017). Given that the task is binary semantic segmentation at the pixel level, binary cross-entropy (BCE) was selected as the loss function, which is well-suited for binary classification problems. The BCE loss is defined as follows:

$$L(y, \hat{y}) = -\frac{1}{N}\sum_{i=1}^{N}\left[y_i log\left(\hat{y}_i\right) + (1 - y_i) log\left(1 - \hat{y}_i\right)\right] \tag{3.4}$$

where N is the number of categories, y is the ground truth label (0 or 1), and $p(y)$ is the predicted probability of label y.

All networks in this study were trained from scratch, without the use of pre-trained weights. During training, the Area Under the Curve (AUC) was monitored as a performance metric.

3.1.3 Landslide Detection Results

To comprehensively evaluate the performance of the proposed model, we conducted a series of experiments within a consistent experimental setting and compared the results of three models: U-Net, U-Net Backbone, and the proposed channel attention-enhanced U-Net. Evaluation metrics derived from the test set were employed to quantify model performance, followed by a multi-scale visual and quantitative mapping of landslide detection accuracy across the study area.

3.1.4 Detailed Landslide Detection Outcomes

The landslide boundaries within the main urban area of Lanzhou were delineated using the aforementioned models. Representative fine-scale identification results are illustrated in Fig. 3.5. In this figure, the first column shows Sentinel-2A false-color composites (bands 2, 3, and 4), the second column displays ground-truth landslide annotations, and the third to fifth columns correspond to the outputs of U-Net, U-Net Backbone, and the proposed attention-augmented U-Net, respectively. Visual inspection reveals that all three models successfully capture landslide boundary information across the selected test scenes.

Closer analysis of the results reveals that the standard U-Net model demonstrates finer delineation of landslide boundaries. This can be attributed to its skip connection design, which effectively fuses low- and high-level features, thereby enriching the semantic representation in the decoding phase. However, despite these advantages, U-Net also presents false positives (as marked in red circles in rows 1–5 of Fig. 3.5c), likely due to insufficient attention to redundant feature channels generated during skip connection fusion. Consequently, small non-landslide objects with spectral similarities to landslides are sometimes misclassified, impairing overall accuracy.

In contrast, the U-Net Backbone model—lacking skip connections—exhibits weaker feature extraction capabilities. This results in coarser boundary definitions and merging of adjacent landslides into single segments, especially in the case of small or narrow landslides (see red circles in rows 1, 4, 5, and 6 of Fig. 3.5d). The absence of hierarchical feature integration leads to insufficient spatial discrimination and a higher rate of over-segmentation.

The proposed model, integrating a channel attention mechanism (SE-block) into the U-Net architecture, achieves superior performance across all representative test areas (Fig. 3.5e). The predicted landslide boundaries closely match the ground-truth data, with minimal false detections, particularly for small-scale and elongated landslides (red circles in rows 2–6). By embedding attention modules into the skip connections, the model effectively suppresses noise, highlights discriminative channels, and focuses on landslide-relevant features. This not only enhances feature learning

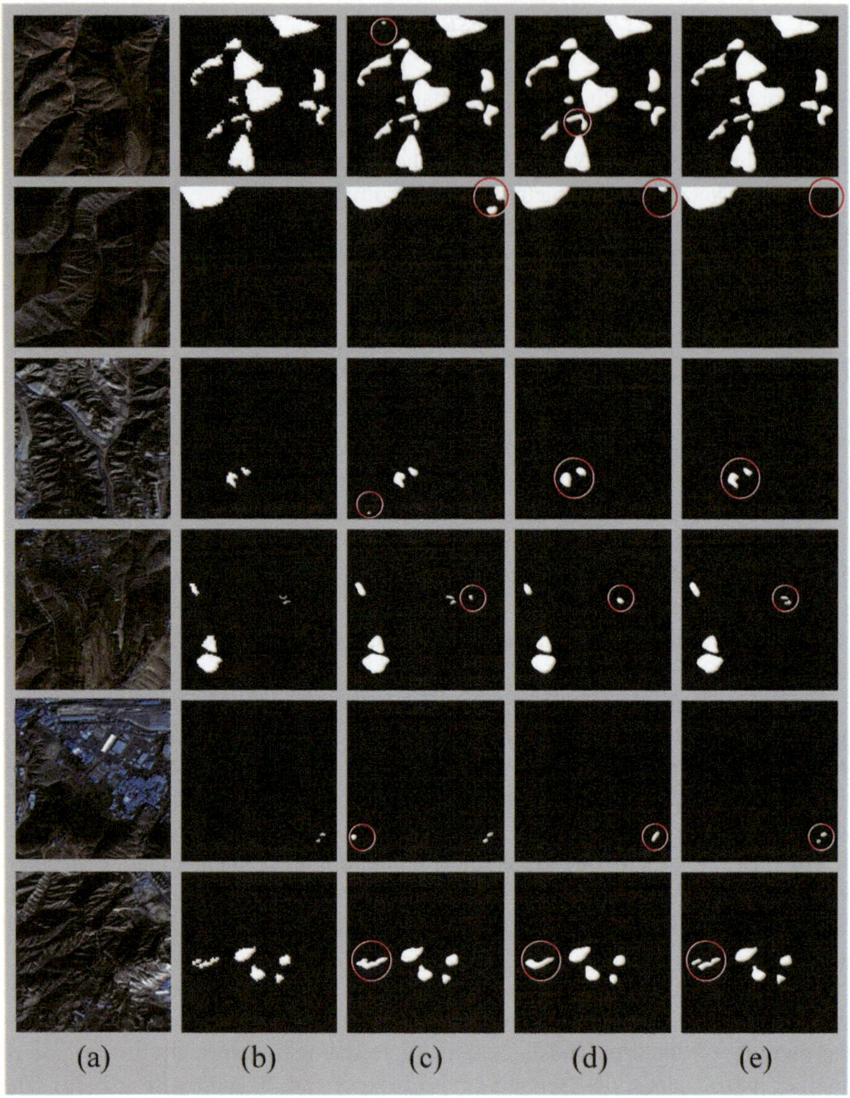

Fig. 3.5 Test results of different models on the test set: **a** landslide image, **b** landslide sample, **c** U-Net results, **d** U-Net Backbone results, **e** Channel attention mechanism U-Net results, reprinted from Ref. Chen et al. (2023), copyright 2025, with permission from Taylor & Francis

and discrimination but also improves model robustness and yields more precise segmentation contours with reduced false positives.

To further evaluate the boundary delineation capability of the proposed model, probability maps were generated to visualize pixel-level classification confidence across three models: U-Net, U-Net Backbone, and the proposed channel attention-enhanced U-Net (Fig. 3.6). These maps depict the raw probability (ranging from 0 to 1) of each pixel being classified as a landslide, with the accompanying color bar indicating the probability scale—blue denoting low confidence and red denoting high confidence. The first column presents the ground truth labels, where each pixel is marked as either landslide or background. Columns two through four illustrate the predicted probability maps from the U-Net, U-Net Backbone, and the improved U-Net with channel attention mechanism, respectively.

From the probability maps, it can be observed that, compared with the ground truth samples (Fig. 3.6a), the proposed channel attention-based U-Net model shows the best pixel classification performance on the test images, particularly achieving finer and more accurate boundary identification for small-scale landslides. In contrast, the U-Net model shows more misclassifications of landslide pixels (Fig. 3.6b), where some non-landslide pixels are assigned a high probability of being landslides, indicating limited ability in learning landslide features and weaker model performance. In the probability maps, the color transitions from low to high probability mainly occur near the landslide boundaries. The U-Net backbone model exhibits more uncertainty at the landslide edges, as the color transition region between blue and red is larger than that in the improved channel attention-based U-Net model (Fig. 3.6c versus 3.6d), indicating lower pixel probabilities at landslide boundaries. Furthermore,

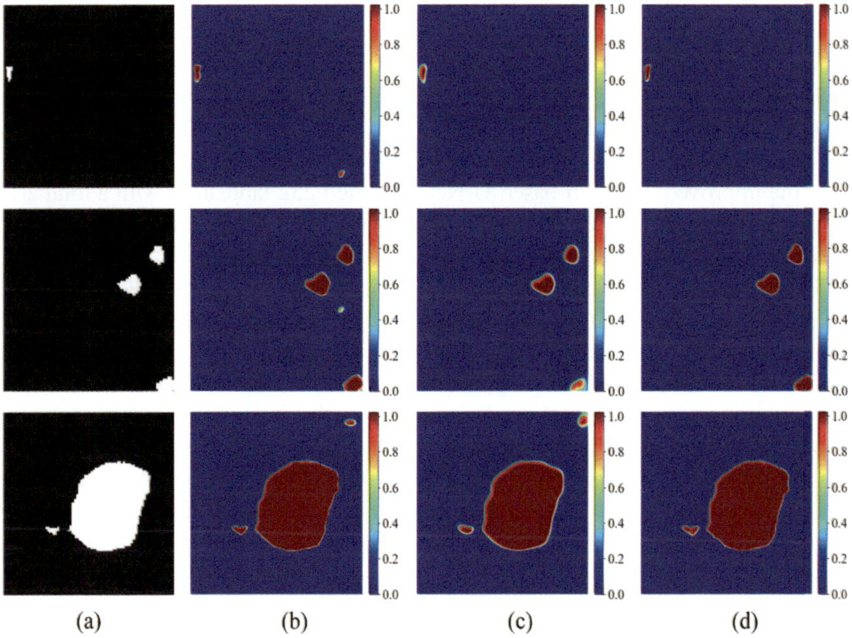

Fig. 3.6 Probability maps of landslide recognition by different models: **a** landslide sample, **b** U-Net results, **c** U-Net backbone results, **d** channel attention mechanism U-Net results, reprinted from Ref. Chen et al. (2023), copyright 2025, with permission from Taylor & Francis

misclassification of non-landslide pixels is observed in the third row in Fig. 3.6c, suggesting that the U-Net Backbone model lacks sufficient precision in landslide boundary delineation, especially for small-scale landslides. In conclusion, the probability maps demonstrate that the improved channel attention-based U-Net model effectively captures landslide features, extracts fine boundary details, and achieves more accurate identification of small-scale landslides with reduced false detection rates.

3.1.4.1 Large-Scale Landslide Detection Outcomes

Landslides in Lanzhou are primarily characterized by mixed-type, mid-level, and small-scale loess slides. The landslide bodies are mainly composed of primary and secondary loess deposits, whose spectral and textural characteristics closely resemble those of the surrounding terrain, making accurate detection particularly challenging. Leveraging Sentinel-2A multi-spectral imagery, this study applied three deep learning models—U-Net, U-Net backbone, and an improved U-Net integrated with a channel attention mechanism—to identify landslides within the central urban area of Lanzhou. Figure 3.7 presents the recognition results, including visualizations of the detected landslide regions and statistical summaries of landslide counts and areas.

Among the three models, the proposed channel attention-enhanced U-Net achieved the most accurate boundary delineation and demonstrated superior performance in detecting small-scale landslides, with results closely matching the ground truth. Quantitative analysis revealed that the reference dataset contains 255 landslides covering a total area of 17.83 km^2. The original U-Net identified 307 landslides (20.00 km^2), while the U-Net Backbone detected 230 landslides (18.88 km^2). In comparison, the improved attention-based U-Net identified 252 landslides with a total area of 19.16 km^2—striking a balance between sensitivity and specificity and aligning most closely with the reference.

These results indicate that the U-Net backbone underperformed in feature learning, leading to higher omission rates, while the original U-Net tended to overestimate landslides, particularly small-scale ones, resulting in more false positives. The proposed attention-based U-Net demonstrated the most reliable recognition accuracy across varying test scenarios. Visually and statistically, it produced results closest to the actual distribution of landslides.

Overall, the superior performance of the proposed model can be attributed to the integration of the SENet channel attention mechanism, which adaptively reweights feature channels after the fusion of low- and high-level representations via skip connections. This allows the network to prioritize channels that contribute more significantly to landslide discrimination, enhancing the model's ability to distinguish between landslide and non-landslide regions. Consequently, it improves the precision of landslide identification while reducing both false positives and false negatives, particularly in large-scale detection scenarios.

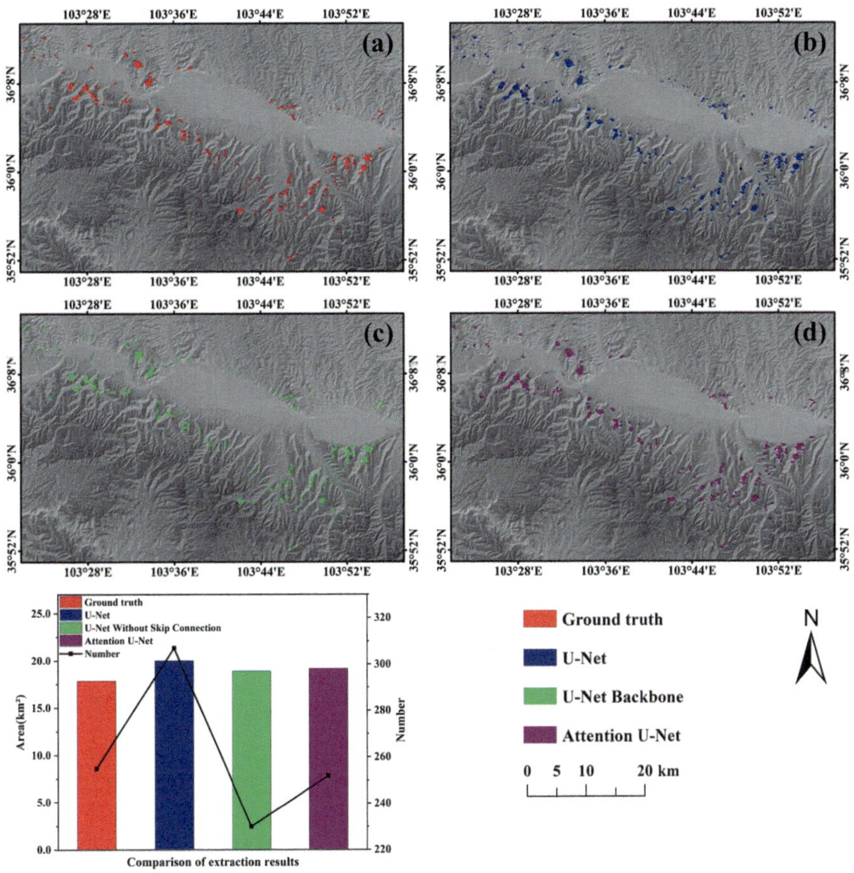

Fig. 3.7 Overall landslide recognition results for different models in the experimental scene, reprinted from Ref. Chen et al. (2023), copyright 2025, with permission from Taylor & Francis

3.1.5 Evaluation of Model Performance

3.1.5.1 Analysis of the Training Process

Under identical experimental settings, landslide extraction models based on U-Net, U-Net Backbone, and the proposed channel attention-enhanced U-Net were trained. The loss function and performance metric curves for both training and validation sets are illustrated in Fig. 3.8. As shown, all three models exhibit stable loss and metric trends, indicating well-tuned training parameters. The area under the ROC curve (AUC) for each model exceeded 90%, demonstrating robust classification performance across the dataset.

The models were trained over 128 epochs. In the early phase, all models exhibited low AUC values, with a rapid increase between epochs 10 and 30. The U-Net model

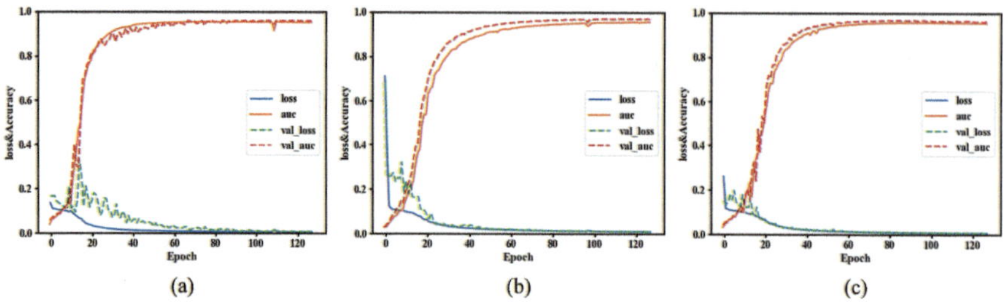

Fig. 3.8 Accuracy and loss curves of the training process for different models: **a** U-Net, **b** U-Net backbone, **c** channel attention mechanism U-Net, reprinted from Ref. Chen et al. (2023), copyright 2025, with permission from Taylor & Francis

converged around epoch 55 (Fig. 3.8a), while U-Net backbone and the proposed channel attention U-Net reached convergence around epochs 70 (Fig. 3.8b) and 60 (Fig. 3.8c), respectively. Among them, the U-Net backbone showed the slowest convergence and largest loss, likely due to limited capacity in learning discriminative landslide features. Moreover, the proposed model showed smoother convergence with smaller loss values, particularly during early training, suggesting improved training stability.

These results suggest that the proposed attention-augmented U-Net outperforms the baseline models in both convergence speed and training efficiency.

3.1.5.2 Comparison of Evaluation Metrics

To quantitatively evaluate detection performance, five key metrics—Precision, Recall, F1-score, MeanIoU, and Kappa—were computed based on confusion matrices (Fig. 3.9). The proposed model outperformed the baseline U-Net and U-Net backbone across all metrics except Recall, achieving superior values in Precision, F1-score, Kappa, and MeanIoU. The Recall of the proposed model (95.19%) was slightly lower than that of U-Net (96.09%), which may stem from U-Net's tendency to misclassify non-landslide objects as landslides, particularly in regions dominated by small-scale landslides. Therefore, while Recall was higher, it may not represent better accuracy. A holistic view of all five indicators demonstrates the superior balance and robustness of the proposed model.

3.1.5.3 Analysis of Model Complexity

Given the importance of rapid and precise landslide detection and emergency management, model efficiency was also assessed using four complexity indicators: parameter count, model size, per-step training time, and inference time (Yi & Zhang, 2020).

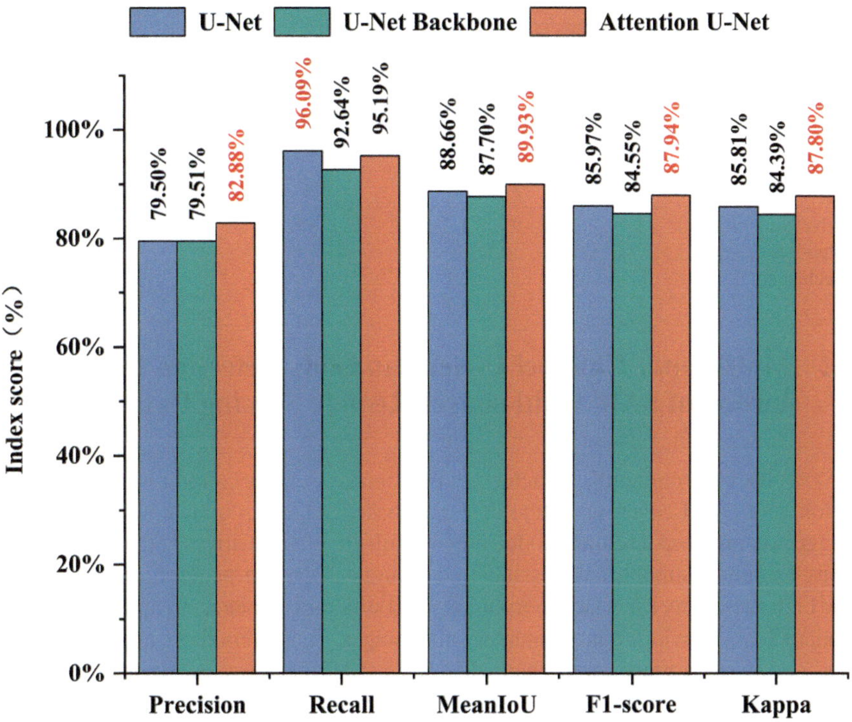

Fig. 3.9 Evaluation metrics for landslide recognition results of different models, reprinted from Ref. Chen et al. (2023), copyright 2025, with permission from Taylor & Francis

As reported in Table 3.1, the proposed model shows a marginal increase in all four metrics compared to baseline models, mainly due to the integration of SENet. However, the additional complexity remains limited, as SENet recalibrates feature channel importance without introducing new spatial dimensions, thereby incurring minimal computational overhead (Niu et al., 2021). Importantly, the attention mechanism effectively suppresses irrelevant information while emphasizing landslide-relevant features, improving model discriminability and reducing both false positives and missed detections.

In summary, the proposed channel attention U-Net demonstrates a favorable trade-off between complexity and performance, making it a compelling model for accurate and efficient landslide detection.

Table 3.1 Comparison of model complexities

Model	Number of parameters (Millions)	Model size (MB)	Training time (Milliseconds per Batch)	Testing time (Seconds)
U-Net	7.76	89.1	677	101
U-Net backbone	6.98	80.1	656	100
U-Net with channel attention mechanism	7.94	91.1	737	120

3.2 Multi-input Channel U-Net Landslide Detection Using Fusion of SAR Multi-source Remote Sensing Data

In complex environments characterized by adverse weather conditions, dense vegetation, and cloud cover, landslide detection based solely on optical remote sensing imagery using deep learning models often results in false positives and missed detections, leading to suboptimal classification accuracy. Given its penetration capabilities and high sensitivity to surface morphology and structural changes, Synthetic Aperture Radar (SAR) provides complementary advantages. To address these challenges, this study proposes a multi-input-channel U-Net model that fuses SAR, optical remote sensing, and topographic data for enhanced landslide detection. The proposed architecture first integrates multi-source data through a tailored U-Net structure, then incorporates an attention mechanism to spatially recalibrate feature maps, emphasizing landslide-relevant patterns. The method is evaluated in a complex terrain setting, and comparative experiments are conducted to demonstrate its effectiveness.

3.2.1 Experimental Setup and Data

The Bailong river basin (BRB), located in southeastern Gansu Province, China, is selected as the test site due to its complex topographic and geological conditions and frequent landslide occurrences. The dataset includes landslide vector labels, SAR imagery (Sentinel-1A), optical imagery (Sentinel-2A), and terrain attributes such as slope and aspect. Landslide vector data are used as pixel-wise ground truth labels for training and validation. Multi-source imagery provides complementary spectral, structural, and topographic information to enhance semantic understanding and support pixel-level classification.

3.2.1.1 Study Area

Geographically, the Bailong river originates from the Gannan Tibetan Autonomous Prefecture and flows through Diebu and Zhouqu counties, and Tanchang, Wudu, and Wen counties in Longnan, Gansu province, Northwest China. The basin spans approximately 381.5 km within Gansu Province, covering an area of around 18,500 km^2 (Zhang, 2018). The region lies within a transition zone from the Qinghai–Tibet Plateau to the Qinling–Daba Mountains and the Loess Plateau (Fig. 3.10), with an annual mean temperature between 2 °C and 15 °C and average annual precipitation near 700 mm. Complex geomorphology, active neotectonics, and the widespread presence of weak rock formations (e.g., phyllite) contribute to high landslide susceptibility. Frequent seismic activity further destabilizes slopes (Meng et al., 2013; Zhang et al., 2016), as evidenced by the 2008 Wenchuan earthquake and subsequent events (e.g., the 2013 Wen County M4.5 and M6.6 earthquakes, and the 2017 Jiuzhaigou M7.0 earthquake), which triggered extensive landslide. The dynamic geological conditions, deeply incised valleys, and tectonic fragmentation create favorable pathways for surface and subsurface water, fostering landslide initiation and progression. As one of China's most landslide-prone regions, the Bailong river basin faces substantial risks to local populations, infrastructure, and broader ecological stability in the upper Yangtze river basin. Given these challenges, reliance on optical data alone is inadequate; instead, an integrated multi-source remote sensing approach is essential for robust and accurate landslide detection.

3.2.2 Data and Preprocessing Procedures

(1) Landslide vector data
A total of 732 landslides were manually delineated through visual interpretation using Google Earth, supported by historical records and field survey validation (Fig. 3.11). The vectorized polygons were rigorously checked and revised to ensure accuracy. These polygons were converted into raster format in ArcMap 10.7 and exported as binary PNG images, where pixel values of 1 and 0 indicate landslide and non-landslide areas, respectively. The resulting label maps were used as ground truth for training the deep learning model.

(2) SAR data
Sentinel-1A SAR data were obtained from the Alaska Satellite Facility (ASF; https://asf.alaska.edu/, accessed on May 6, 2025). Sentinel-1A, launched by the European Space Agency (ESA) in April 2014, offers global radar imaging with a 12-day revisit cycle. Its imaging capabilities are not affected by weather, cloud cover, or illumination, making it well-suited for continuous surface observation under challenging environmental conditions.

In this study, SLC-mode Sentinel-1A images acquired on May 29, 2021, using both VH and VV polarizations in Interferometric Wide (IW) swath mode,

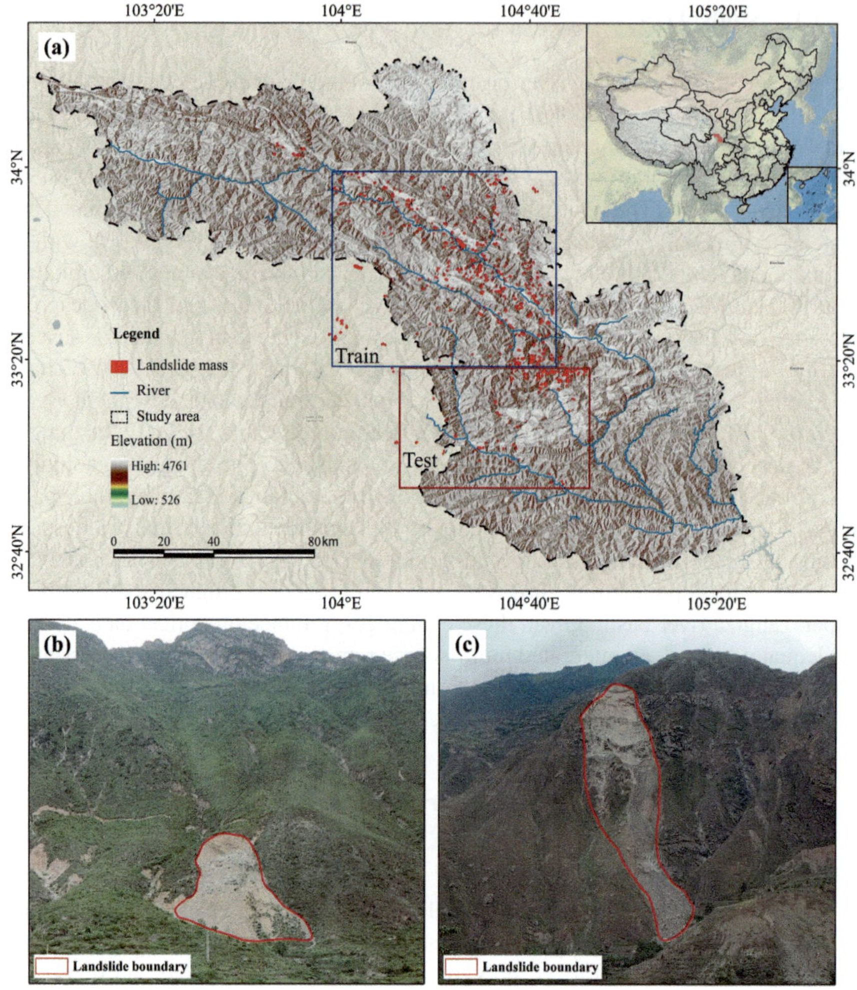

Fig. 3.10 Experimental Scene (Bailong River Basin), reprinted from Ref. Chen et al. (2024), copyright 2025, with permission from IEEE

were selected. Given the speckle noise and low contrast of SAR images, preprocessing was performed using ENVI SARscape software developed by sarmap. The workflow included multilooking, speckle filtering, geocoding, and radiometric calibration (detailed in Tables 3.2, 3.3 and 3.4). The processed images were imported into ArcMap 10.7, resampled, and exported in PNG format for subsequent analysis (Fig. 12a, b).

(3) Optical Data
Cloud-free Sentinel-2A multispectral data acquired on April 29, 2021, were used as the optical data source. The imagery was atmospherically corrected using the

Fig. 3.11 Field survey of the Bailong river basin

Table 3.2 Parameter settings for multi-view

	Key parameters
Multi-view	Time domain
Number of range views	1
Number of Azimuth views	1

Table 3.3 Filtering parameter settings

	Key parameters
Filtering method	Lee
Number of equal views	−1
Azimuthal window size	5
Range window size	5

Table 3.4 Geocoding and radiometric calibration parameter settings

	Key Parameters
Pixel size	15
Radiometric calibration	True
Scattering area	Local incidence angle
Radiometric normalization	True
Local incidence angle correction	False
Output type	Linear

Sen2Cor module in ESA's SNAP software to generate bottom-of-atmosphere (BOA) reflectance products. Subsequently, the images were resampled to 10 m resolution, and 12 spectral bands were extracted. These bands were clipped to the study area using ArcMap 10.7 and exported in PNG format. Representative bands are shown in Fig. 3.12c and d.

(4) Topographic Data
Slope and aspect layers were derived from ALOS 12.5 m DEM data, generated using the Phased Array type L-band Synthetic Aperture Radar (PALSAR) onboard the ALOS satellite. These topographic variables provide essential auxiliary information for landslide detection. The slope and aspect rasters were processed and resampled in ArcMap 10.7, then exported as PNG images (Fig. 3.12e, f).

3.2.2.1 Dataset Creation and Partitioning

To accommodate hardware constraints and streamline the training of deep learning models, the original remote sensing imagery was partitioned, and a dataset was constructed using corresponding landslide sample patches. Considering image resolution and typical landslide scales, each image was divided into 256 × 256 pixel tiles using a sliding window approach with a stride of 128 pixels. Sentinel-1A, Sentinel-2A, slope, aspect, and annotated landslide images were all tiled using this consistent strategy. To enrich sample diversity, reduce overfitting, and improve model generalization and robustness, data augmentation techniques such as mirroring and rotation were applied. To simplify training, image patches that did not contain any landslide

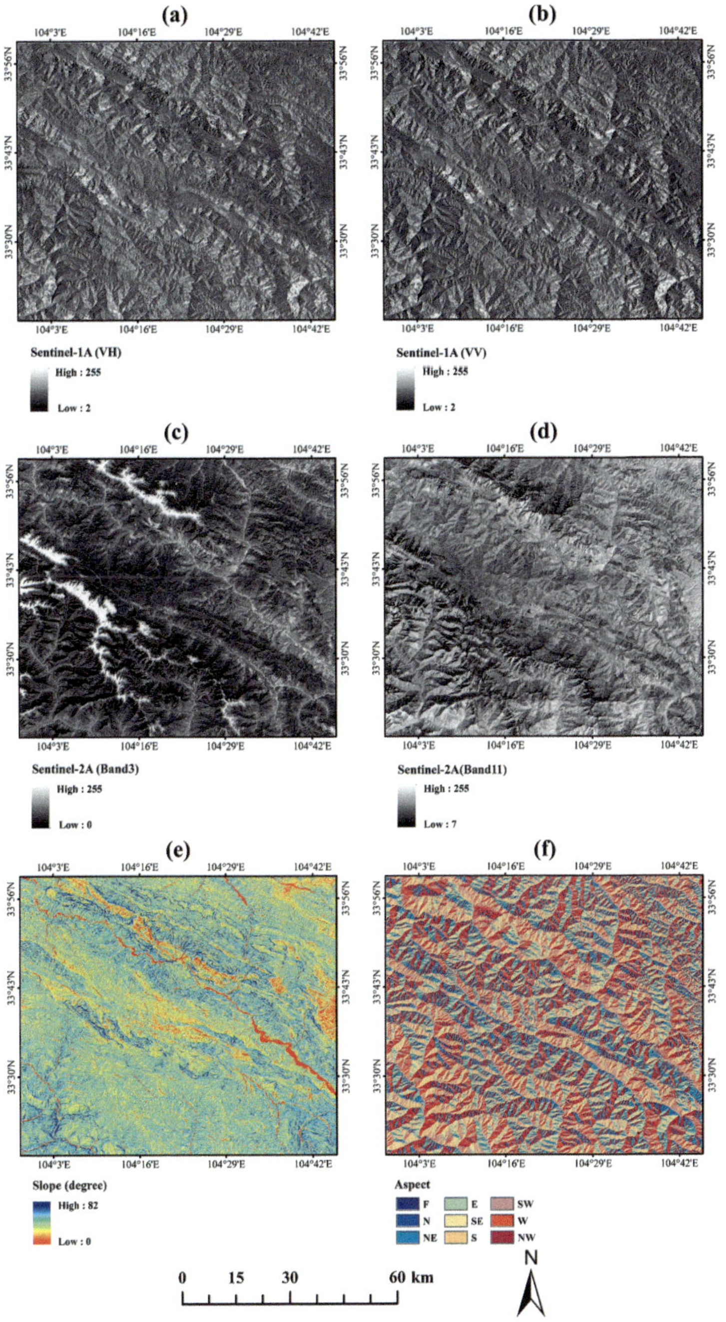

Fig. 3.12 SAR, optical, and terrain image data. **a** Sentinel-1A (VH), **b** Sentinel-2A (VV), **c** Sentinel-2A (Band 3), **d** Sentinel-2A (Band 11), **e** slope, **f** aspect, reprinted from Ref. Chen et al. (2024), copyright 2025, with permission from IEEE

Fig. 3.13 Samples from the partial dataset, reprinted from Ref. Chen et al. (2024), copyright 2025, with permission from IEEE

features were excluded. The final dataset was randomly split into training and validation sets with a ratio of 8:2, ensuring that the two subsets were mutually exclusive (Fig. 3.13).

3.2.3 Construction of the Multi-Input Channel U-Net Model

3.2.3.1 Model Development Approach

This study aims to enhance landslide detection by integrating multi-source SAR and optical remote sensing data. A critical challenge lies in how to effectively structure these data for input into a deep learning model. Moreover, non-landslide background dominates the imagery, potentially reducing classification accuracy. To address this, a multi-input channel design was proposed, treating different data sources as separate input channels to facilitate joint processing. In computer vision, attention mechanisms can selectively emphasize salient spatial features, enabling weighted aggregation across feature maps and promoting more focused learning (Brauwers & Frasincar, 2023). Consequently, instead of the traditional U-Net approach of directly concatenating encoder and decoder features via skip connections, an attention module

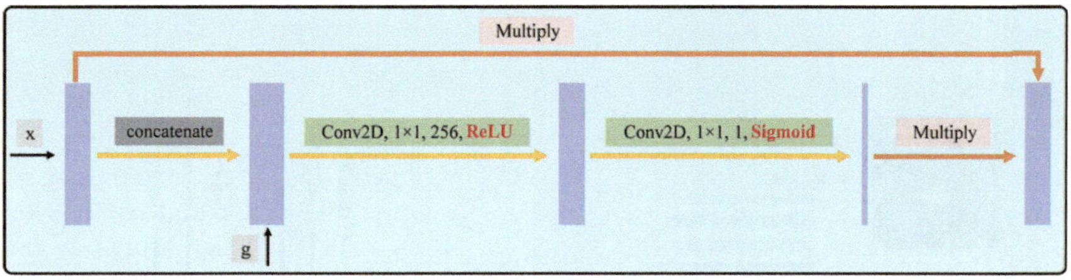

Fig. 3.14 Attention mechanism module, reprinted from Ref. Chen et al. (2024), copyright 2025, with permission from IEEE

was incorporated to refine spatial feature weighting and emphasize landslide-related features (Oktay et al., 2018). This module includes feature concatenation, two successive 1 × 1 convolutions, and a Hadamard product operation. Its architecture is illustrated in Fig. 3.14, and the structure of the multi-channel U-Net is shown in Fig. 3.15.

The model architecture first groups the inputs into two categories—SAR/optical imagery and topographic features (slope and aspect)—which are separately processed through initial convolution-pooling layers. Their resulting feature maps are concatenated along the channel dimension and jointly passed through the subsequent encoding layers. In the attention module, features extracted from the encoder (x) are refined through additional convolution operations, while decoder features (g) of matching spatial dimensions are concatenated with x. The combined feature map then passes through two 1 × 1 convolutions with ReLU and sigmoid activations, respectively. ReLU mitigates overfitting, while Sigmoid produces a single-channel spatial attention map that suppresses irrelevant features. This attention map is element-wise multiplied (Hadamard product) with the encoder feature map to generate attention-weighted features, which are fused with the decoder's upsampling output for accurate spatial reconstruction. This mechanism not only strengthens semantic consistency between encoder and decoder but also overcomes the fixed weighting limitations of conventional U-Net skip connections, thereby improving the model's ability to learn and localize landslide features with greater precision.

3.2.3.2 Model Parameter Configuration and Training Process

To ensure a balance between computational efficiency and hardware constraints, preliminary experiments were conducted, and the number of training iterations was set to 128 with a batch size of 16. The learning rate was fixed at 0.0001, and the Adam optimizer was employed due to its adaptive learning rate mechanism and computational efficiency. As a refinement over Adagrad and RMSprop, Adam is insensitive to gradient scaling and offers ease of implementation. Given that the task involves binary classification at the pixel level, binary cross-entropy was selected as the loss function. The AUC metric was used to monitor training performance. All

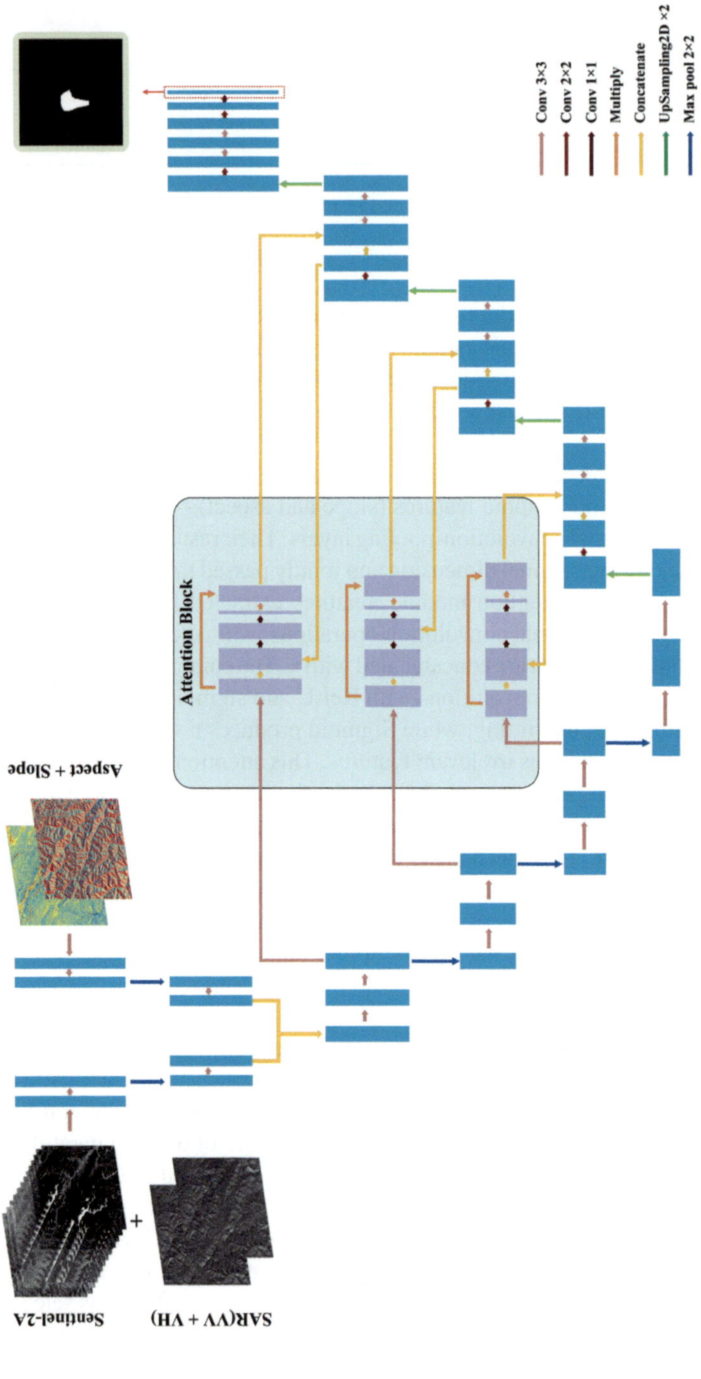

Fig. 3.15 Multi-input channel U-Net architecture, reprinted from Ref. Chen et al. (2024), copyright 2025, with permission from IEEE

deep learning models were developed using the Python programming language and implemented with the TensorFlow API and Keras library.

3.2.4 *Landslide Detection Results*

This section presents the application and comprehensive evaluation of the proposed method in a designated experimental setting. Models were first trained using datasets with various data types within the training area. Subsequently, the trained models were applied to landslide detection in the test area. A comparative analysis from both visual and quantitative perspectives was conducted to evaluate the landslide detection performance of each input data configuration, thereby verifying the effectiveness of integrating multi-source SAR remote sensing data. Additionally, to assess the performance of the multi-input U-Net model with SAR data fusion, we generated TP, FP, and FN maps from the detection results using different network architectures. These visual and comparative experiments offer a thorough evaluation of the proposed method and provide insights for future enhancements in landslide detection methodologies.

3.2.4.1 Landslide Detection Outcomes Using SAR Multi-source Remote Sensing Data

To further verify the usability and effectiveness of the proposed model in fusing multi-source SAR data for landslide detection, four sets of comparative experiments were designed: (1) conventional U-Net with 12-band Sentinel-2A imagery (U-Net + 12 channels); (2) conventional U-Net with fused SAR multi-source remote sensing data, including dual-polarization Sentinel-1A, 12-band Sentinel-2A, slope, and aspect images (U-Net + 6 channels); (3) multi-input U-Net with 12-band Sentinel-2A imagery (multi-input U-Net + 12 channels); and (4) multi-input U-Net with fused SAR multi-source remote sensing data (multi-input U-Net + 16 channels). All models were trained using the same dataset from the training region and tested on the same test area. To provide an overview of the landslide detection results from each method, all predicted image tiles were mosaicked into a complete landslide map of the test area based on geographic coordinates. Furthermore, four quantitative evaluation metrics—Recall, Precision, F1-score, and MIoU—were calculated based on confusion matrices (Figs. 3.16 and 3.17).

By comparing Fig. 3.16a, b and c, d, it can be observed that for both the conventional U-Net and the multi-input U-Net models, the integration of multi-source SAR data generally yields visually better landslide detection results than those using only Sentinel-2A optical imagery. The overall delineation is finer, with fewer false positives and false negatives. Quantitatively, detection results using fused SAR data generally outperform those using optical data alone across all four-evaluation metrics.

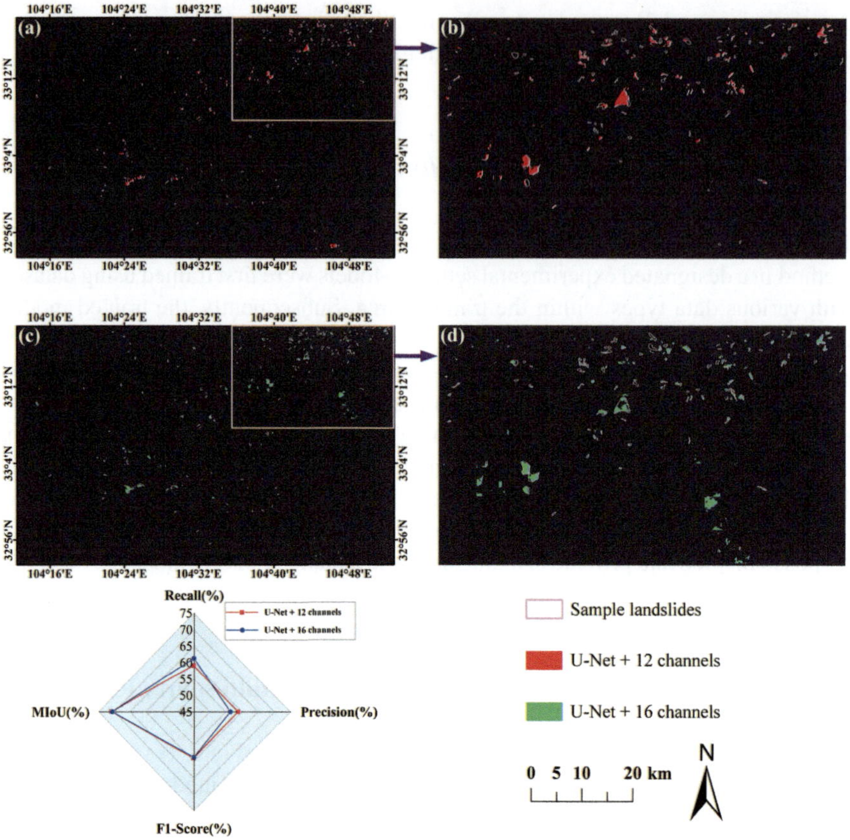

Fig. 3.16 Landslide recognition results of the traditional U-Net model with different data fusion in the test area. **a** Landslide recognition results using traditional U-Net fused with optical remote sensing images, **b** Specific region within (**a**), **c** Landslide recognition results using traditional U-Net fused with SAR multi-source remote sensing data, **d** Specific region within (**c**), reprinted from Ref. Chen et al. (2024), copyright 2025, with permission from IEEE

A high Recall indicates greater coverage, i.e., more landslides were detected regardless of how many non-landslides were mistakenly classified as landslides (e.g. in Fig. 3.16, the multi-input U-Net with SAR fusion yielded a larger landslide area, suggesting more comprehensive detection). High Precision indicates that most of the detected landslides are actual landslides. The F1-score balances Recall and Precision, reflecting overall model performance. MIoU evaluates pixel-level similarity between prediction and ground truth. In this study, since the number of background pixels (non-landslides) significantly exceeds that of landslides, MIoU values in all four experiments were higher than Recall, Precision, and F1-score. For the multi-input U-Net, the F1-score of the SAR-fused model was approximately 3.5% higher than that of the optical-only model, confirming that incorporating SAR multi-source

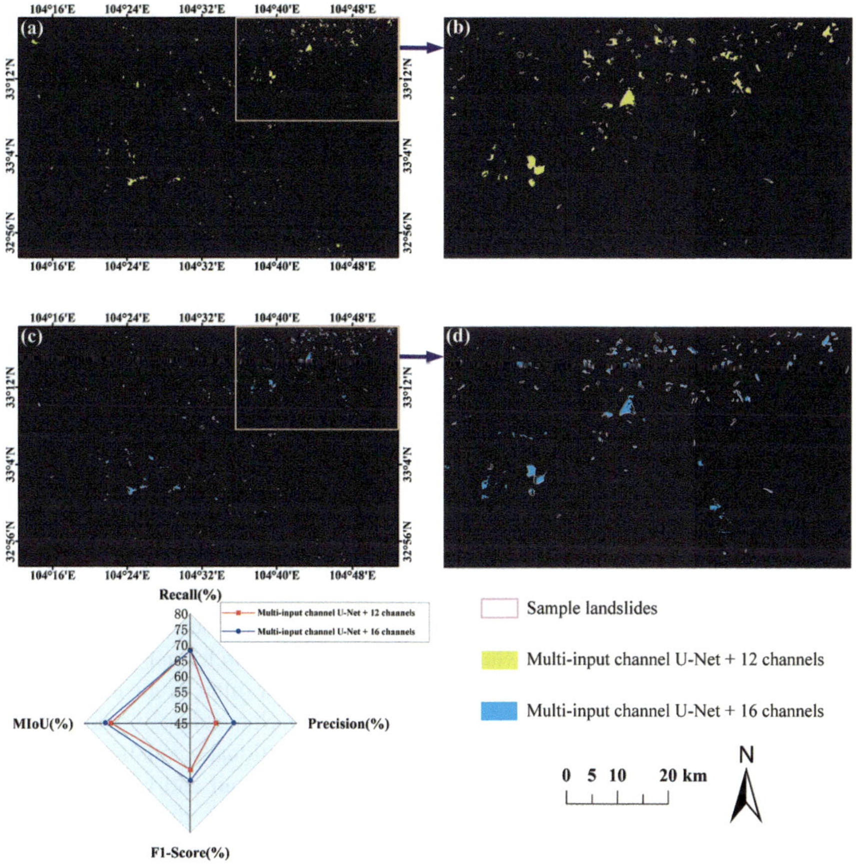

Fig. 3.17 Landslide recognition results of the multi-input channel U-Net model with different data fusion in the test area. **a** landslide recognition results using multi-input U-Net fused with optical remote sensing images, **b** specific region within (**a**), **c** landslide recognition results using multi-input U-Net fused with SAR multi-source remote sensing data, **d** specific region within (**c**), reprinted from Ref. Chen et al. (2024), copyright 2025, with permission from IEEE

data improves the model's robustness in complex scenarios by effectively reducing false positives and false negatives.

Comparison between different models reveals that, due to the complex geographical environment of the Bailong river basin—characterized by rugged terrain, variable climate, dense vegetation cover, and frequent cloud and fog—relying solely on single-source optical remote sensing imagery is insufficient to fully extract latent landslide-related features. Such limitations hinder the model's ability to learn from comprehensive landslide characteristics embedded in the imagery. In contrast, landslide detection methods that integrate multi-source SAR remote sensing data effectively introduce and leverage the rich and diverse information across modalities. These fused multi-dimensional features enhance the contextual understanding of surface objects in complex environments, allowing the model to learn landslide features from multiple perspectives and distinguish landslides from surrounding similar terrain types. Both visual and quantitative evaluations confirm the usability and effectiveness of incorporating SAR-based multi-source remote sensing data in landslide detection.

3.2.4.2 Accuracy Analysis of the Model

In this study, a multi-input U-Net deep learning model was developed for landslide detection by integrating multi-source remote sensing data, including SAR (Sentinel-1A), optical (Sentinel-2A), and topographic features (slope and aspect). To verify the effectiveness and superiority of the proposed model, its landslide detection performance was compared with that of a traditional U-Net model in the test area. The results indicate that both models can identify the majority of landslides within the test region. To provide a clearer visual comparison, several subregions were randomly selected and enlarged for display (Fig. 3.18).

By comparing the actual landslide samples with the detection results from both the multi-input U-Net and traditional U-Net models, it is observed that while the traditional U-Net model can capture relatively complete landslide boundaries, it exhibits a higher number of false positives (highlighted with yellow circles in Fig. 3.19c) and false negatives (red circles in Fig. 3.19c). False positives often occur near known landslide areas and are generally of small spatial extent. Some closely spaced landslides are incorrectly connected at their boundaries. Additionally, small-scale landslides are prone to being missed, indicating insufficient detail and weaker feature learning capabilities in the traditional U-Net model. In contrast, the multi-input U-Net model produces more accurate and complete detection results that are closer to the reference landslide data (Fig. 3.19d), with fewer false positives and false negatives. This demonstrates the comprehensive capability of the multi-input U-Net model in learning landslide features and distinguishing landslides from non-landslides more effectively, thereby enhancing overall detection accuracy.

To further evaluate the precision of the detection results, true positives (TP), false positives (FP), and false negatives (FN) were calculated based on actual landslide samples and detection results for both the traditional U-Net and multi-input U-Net models. These results were then visualized for comparison at a broader regional scale (Fig. 3.20). As shown in Fig. 3.20, compared with the traditional U-Net (Fig. 3.20a, b), the multi-input U-Net model yields more TP pixels (blue regions in Fig. 3.20c, d: predicted as landslide and actually landslide), and fewer FP (red regions: predicted as landslide but actually non-landslide) and FN pixels (yellow regions: predicted as non-landslide but actually landslide). This suggests that the multi-input U-Net model achieves a better balance between Precision and Recall, enabling more accurate landslide identification with fewer misclassifications, thus validating the effectiveness of the proposed method. In addition, several subregions were randomly selected to visualize and enlarge the TP, FP, and FN results (Fig. 3.20), allowing a multi-scale

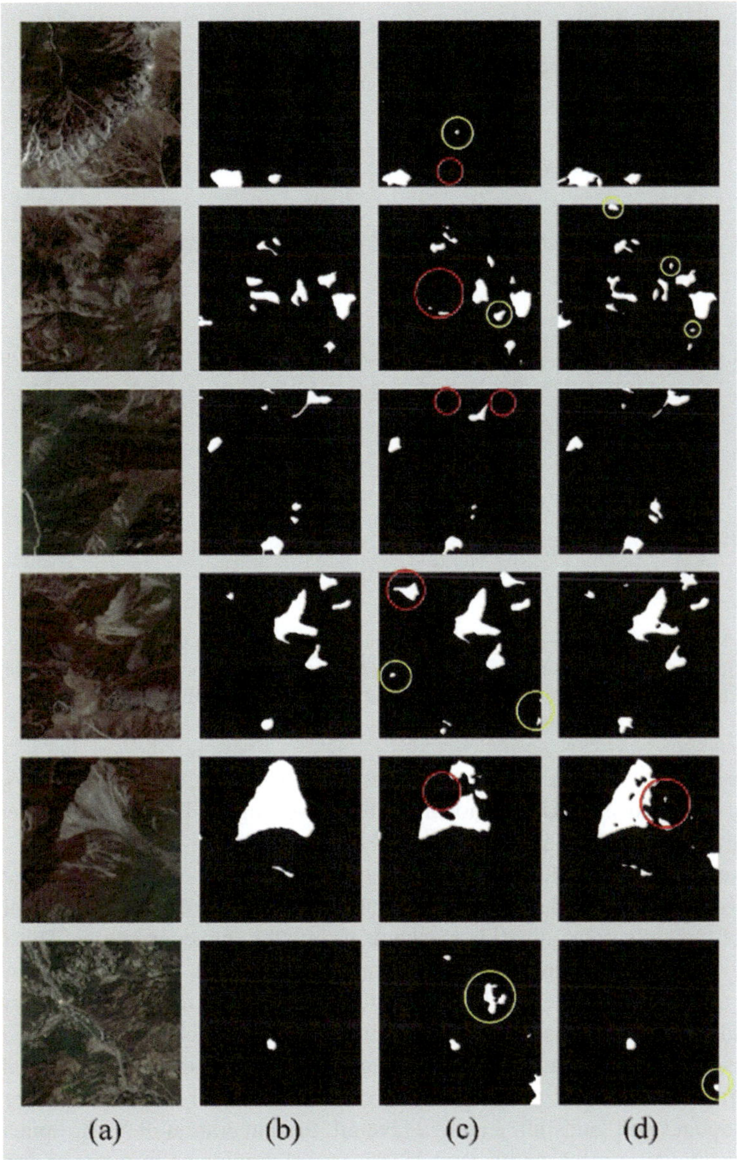

Fig. 3.18 Partial landslide recognition results of different models fused with SAR multi-source remote sensing data in the test area. **a** Optical (Sentinel-2A) image, **b** ground truth landslide sample, **c** landslide recognition result using U-Net, **d** landslide recognition result using multi-input channel U-Net, reprinted from Ref. Chen et al. (2024), copyright 2025, with permission from IEEE

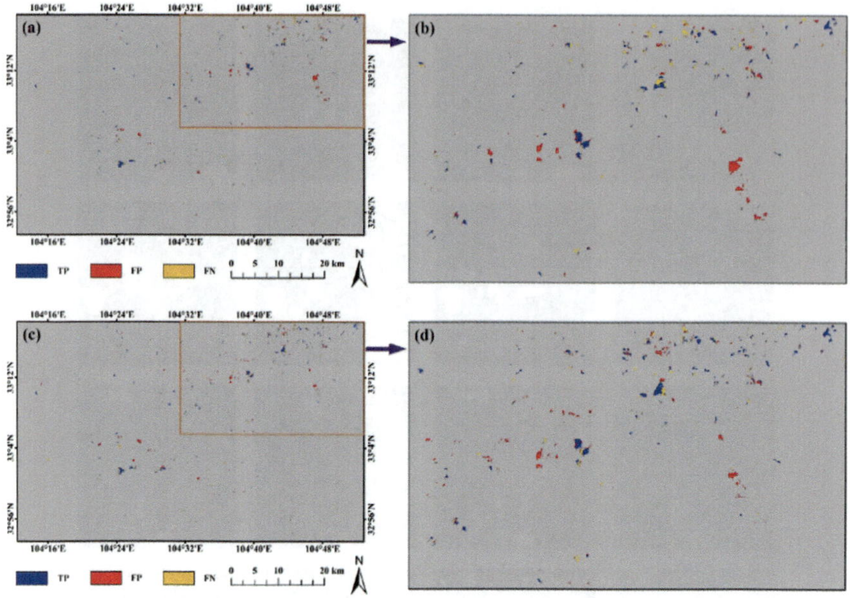

Fig. 3.19 Overall TP, FP, and FN maps of the landslide recognition results. **a** Result using U-Net, **b** specific region of (**a**), **c** result using multi-input channel U-Net, **d** specific region of (**c**), reprinted from Ref. Chen et al. (2024), copyright 2025, with permission from IEEE

comparison of detection accuracy. The results consistently demonstrate the superiority of the multi-input U-Net model, as it achieves higher TP and lower FP and FN compared to the traditional U-Net model.

In summary, the multi-input U-Net model constructed in this study effectively incorporates SAR-based multi-source remote sensing data through additional input channels and integrates attention mechanisms into the skip connections. This allows spatial weights of feature maps to be adaptively adjusted, improving the model's ability to process complex background information while emphasizing and learning landslide-related features. By enhancing feature fusion across encoding and decoding stages and addressing the fixed spatial weight limitations and semantic gaps inherent in the traditional U-Net, the proposed model enables better differentiation between landslide and non-landslide features. Overall, the integration of SAR, optical, and topographic data in the multi-input U-Net model significantly improves landslide detection performance.

3.2.5 Model Performance Evaluation

Grad-CAM, a gradient-based class activation mapping method, offers a visual interpretation of deep network behavior by highlighting model attention areas through

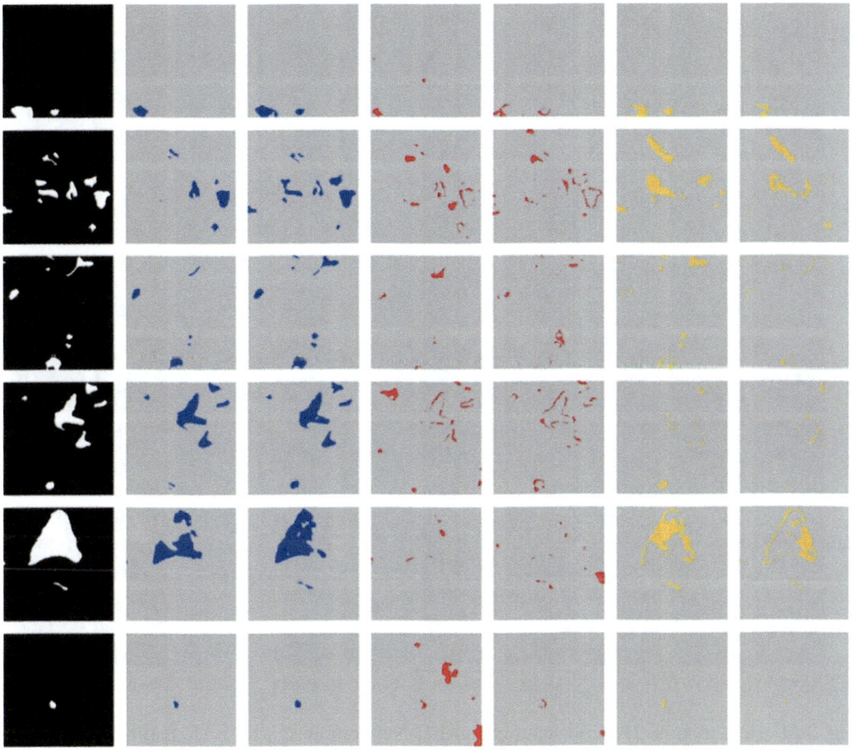

Fig. 3.20 TP, FP, and FN maps of the landslide recognition result in typical regions. **a** Real landslide sample, **b** TP using U-Net, **c** TP using multi-input channel U-Net, **d** FP using U-Net, **e** FP using multi-input channel U-Net, **f** FN using U-Net, **g** FN using multi-input channel U-Net, reprinted from Ref. Chen et al. (2024), copyright 2025, with permission from IEEE

heatmaps. This technique enhances model interpretability by visually illustrating decision rationales and highlighting critical features contributing to the final prediction (Hacıefendioğlu et al., 2021; Selvaraju et al., 2017). Prior to training the multi-input-channel U-Net model, all data sources were abstracted into a unified 16-band input, allowing simultaneous integration and learning across diverse modalities. To improve the interpretability of the proposed method, Grad-CAM was applied to each input channel to generate individual heatmaps, which were overlaid on the respective band images to visualize the spatial regions influencing the model's decisions. The results are presented in Fig. 3.21.

As illustrated, red regions denote areas receiving higher model attention, with darker hues indicating stronger contribution to landslide classification. These regions are interpreted as highly relevant for landslide identification. In contrast, blue areas signify lower attention, indicating limited relevance to the classification task (Panwar

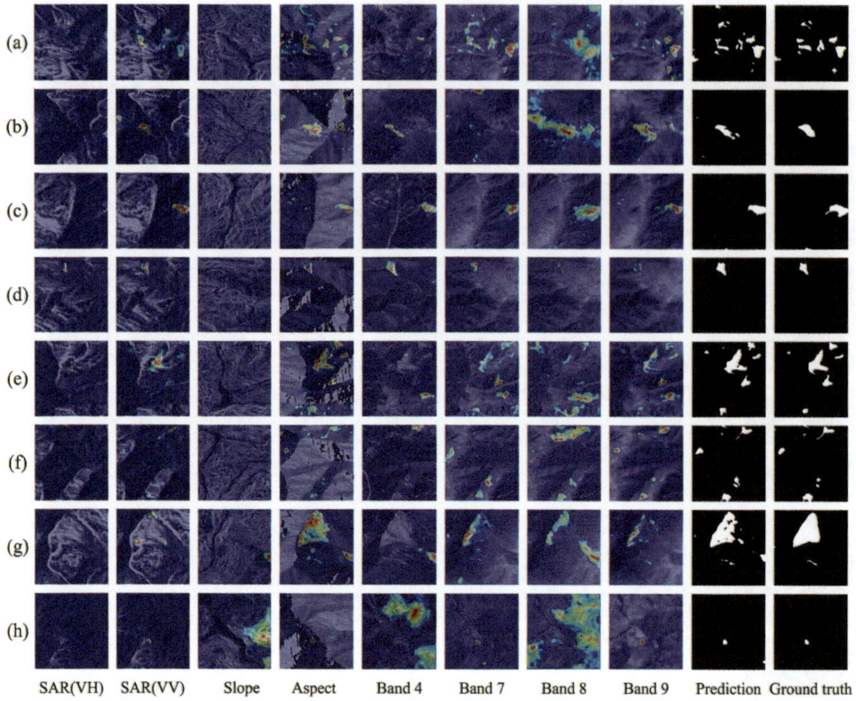

Fig. 3.21 Heat map of 21 multi-input channel U-Net combined with SAR multi-source remote sensing data for landslide identification, reprinted from Ref. Chen et al. (2024), copyright 2025, with permission from IEEE

et al., 2020). These visualizations demonstrate that each input band contributes uniquely to the recognition process.

The study leverages multi-source SAR-based remote sensing data for landslide detection. Owing to intrinsic differences in physical characteristics, resolution, and sensor parameters, multi-source data offer heterogeneous and complementary information. SAR, in particular, provides significant advantages due to its ability to penetrate vegetation and cloud cover, which is crucial for identifying landslides potentially obscured by such features (Metternicht et al., 2005; Lissak et al., 2020). Furthermore, SAR is sensitive to surface and structural changes—key indicators of landslide events (Yoshimatsu & Abe, 2006). This study utilizes dual-polarization Sentinel-1A SAR data (VH and VV), where VH denotes vertical transmission and horizontal reception, while VV denotes both vertical transmission and reception. The differential backscattering characteristics under varying polarization enhance feature extraction. The integration of multiple polarizations provides complementary structural and textural information, thereby improving landslide delineation, reducing both false alarms and omissions, and enhancing the overall classification performance (Figs. 3.19 and 3.20).

As shown in Fig. 3.21, the model's response to landslide features in certain optical bands was relatively coarse, while SAR imagery elicited more focused and precise attention (e.g. Fig. 3.21a, c, h). In particular, Fig. 3.21h reveals spurious attention in some optical bands (e.g., Bands 4 and 8), whereas SAR bands display cleaner and more accurate focus areas. Additionally, slope and aspect data contribute valuable terrain-related information that optical imagery alone cannot capture. Given the strong association between landslides and terrain steepness and orientation, incorporating these topographic variables enables the model to better capture geomorphological characteristics relevant to landslide detection (Fig. 3.21e, g, h).

The proposed multi-input-channel U-Net architecture, enhanced with an attention mechanism (Fig. 3.14), is designed to handle heterogeneous inputs effectively. By assigning adaptive weights to different data types, the network selectively emphasizes landslide-related information across modalities. The attention module addresses the limitation of fixed spatial weighting in conventional U-Net feature fusion, allowing the model to focus on high-contribution areas (e.g., the darker/red regions in Fig. 3.21). This targeted focus improves the model's capability in both landslide localization and classification. Through effective integration of multi-source remote sensing data, the model achieves greater accuracy, robustness, and interpretability in landslide detection.

3.3 Multi-model Ensemble Landslide Detection Considering Spectral, Terrain, and Texture Features

This chapter presents a landslide detection approach that integrates spectral indices, topographic variables, and texture features derived from optical remote sensing imagery. To enhance the robustness and generalization capability of the model, an ensemble learning strategy is employed by combining multiple deep learning architectures. This integrated framework mitigates the limitations of individual models and enables more accurate and intelligent pixel-level landslide identification.

3.3.1 Study Area and Data Sources

Given the complex terrain and high frequency of landslide events in the Bailong River Basin, the use of optical remote sensing data alone poses limitations for accurate landslide detection. Therefore, this chapter selects the Bailong river basin (Fig. 3.10) as the experimental area to validate the effectiveness of the proposed method. The geographic and environmental context of the study area is detailed in Section 3.2.1.

The dataset consists of landslide vector annotations, optical remote sensing imagery, spectral indices, topographic variables, and texture measures. The annotated landslide vectors, after preprocessing, provide pixel-level ground-truth labels for model training and validation. The optical, spectral, topographic, and texture features offer essential semantic cues for pixel-wise classification by deep learning models.

3.3.1.1 Data Description and Preprocessing

(1) Landslide Vector Labels
The landslide vector data are derived from Sect. 3.2 and were updated through field validation to enhance accuracy. After preprocessing, the resulting dataset provides reliable reference samples for training the proposed deep learning models.
(2) Sentinel-2A data
Sentinel-2A data provided by the ESA were used as optical remote sensing data. Cloud-free Sentinel-2A data from April 29, 2021, were selected. After preprocessing, 12 bands were obtained. To enhance data usability, provide more detailed surface information, and improve landslide feature and boundary recognition while reducing computational complexity and cost, four 10-m resolution bands (Bands 2, 3, 4, and 8) were selected. The resampled bands were imported into ArcMap 10.7, clipped to the study area, and exported in PNG format. Examples of these bands are shown in Fig. 3.22a–d.
(3) Spectral Indices
Spectral indices derived from optical imagery can capture post-landslide phenomena such as soil exposure, vegetation loss, and moisture variation. In this study, four indices were computed: Normalized Red Band (NR), Normalized Green Band (NG), Normalized Blue Band (NB), and the Normalized Green–Red Difference Index (NGRDI) (Lou et al., 2023). These indices were calculated using predefined formulas based on relevant Sentinel-2A bands. The outputs were resampled in ArcMap 10.7 and exported as PNG files (Fig. 3.22e–h).

NR:

$$NR = \frac{R}{B+G+R} \tag{3.5}$$

NG:

$$NG = \frac{G}{B+G+R} \tag{3.6}$$

NB:

$$NNB = \frac{B}{B+G+R} \tag{3.7}$$

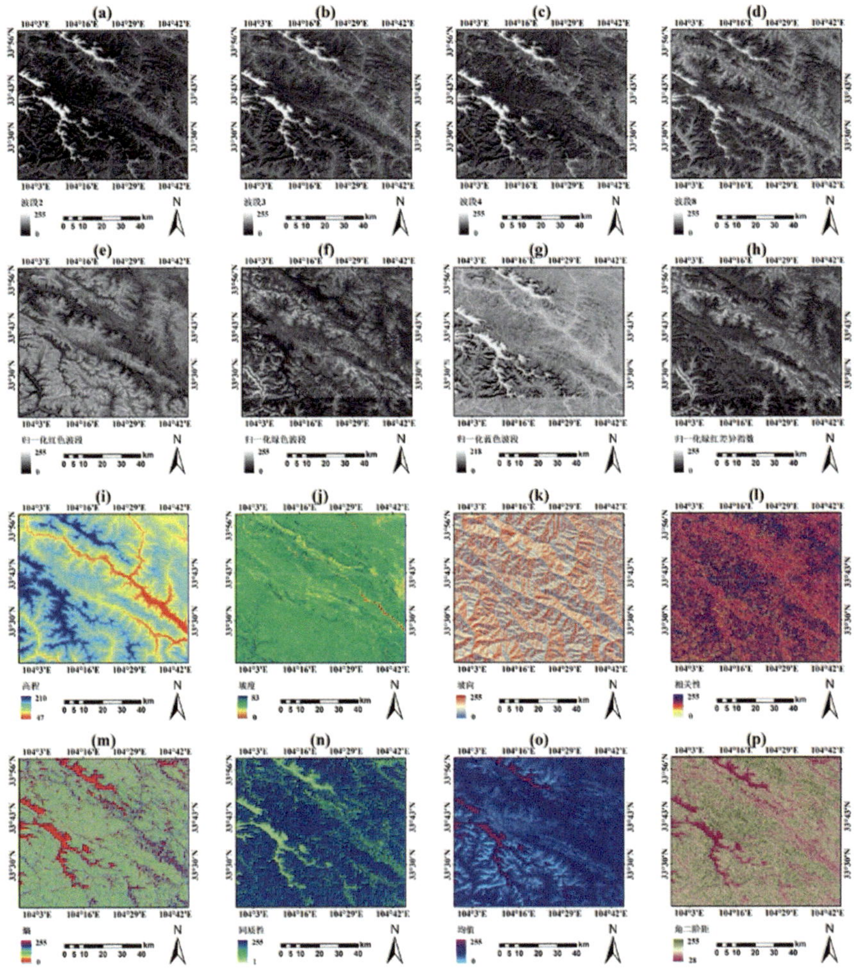

Fig. 3.22 Optical imagery, spectral index, terrain factor and texture index data

NGRDI:

$$NGRDI = \frac{G \cdot R}{G + R} \tag{3.8}$$

where R, G, and B denote the red, green, and blue spectral bands, respectively.

(4) Topographic Features

Topographic factors are crucial parameters representing terrain features, significantly influencing the formation and distribution of landslides. For instance, steeper slopes are more prone to landslides, and aspect can also affect landslide distribution. The topographic data used in this study include DEM, slope, and aspect. ALOS 12.5 m

DEM data were used and resampled to match the resolution of the slope and aspect data, and exported in PNG format. Slope and aspect were derived from the DEM data and similarly resampled and exported (Fig. 3.22i–k).

(5) Texture Features

Texture measures describe spatial heterogeneity in surface features, which often change significantly due to landslide activity. Five texture indices—Correlation, Entropy, Homogeneity, Mean, and Angular Second Moment—were computed based on Gray Level Co-occurrence Matrix (GLCM) techniques using Sentinel-2A data in ENVI 5.6 (Lou et al., 2023). The texture maps were resampled and exported in PNG format (Fig. 3.22l–p).

3.3.1.2 Dataset Creation and Partitioning

Given the large size of the original remote sensing imagery and the limitations of available GPU memory, the raw images could not be directly fed into the model for training. To address this issue, the images were divided into uniformly sized sub-images. The choice of patch size is critical: smaller patches may fail to capture large-scale landslides, whereas larger ones risk exceeding memory capacity. Based on the spatial scale of landslides in the study area and hardware constraints, a patch size of 256 × 256 pixels was selected. A sliding window approach with a stride of 128 pixels was employed to generate overlapping tiles. All input layers—including optical imagery, spectral indices, topographic factors, texture metrics, and corresponding ground-truth labels—were partitioned using the same strategy. To ensure effective feature learning, only patches containing at least one landslide region were retained.

To further augment the dataset and enhance the model's generalization capability across varying data distributions, data augmentation techniques were applied (Yu et al., 2017). Specifically, mirror flipping and 90° counterclockwise rotations were performed on the image-label pairs, resulting in a total of 3,153 samples. The dataset was then randomly split into a training set (2,850 samples) and a validation set (573 samples) at a ratio of 8:2. These subsets were mutually exclusive (Fig. 3.23). The training set was used to optimize model weights, while the validation set was used to evaluate model performance and select optimal hyperparameters (Lyons et al., 2018). This approach enabled full utilization of the available data, improving model robustness and predictive performance.

The study area was spatially partitioned into independent training and testing zones, ensuring no data leakage between them. All training samples were derived solely from the training zone, and no information from the testing zone was used during model training. This strict spatial separation allows an objective evaluation of the model's performance and generalization ability under real-world conditions.

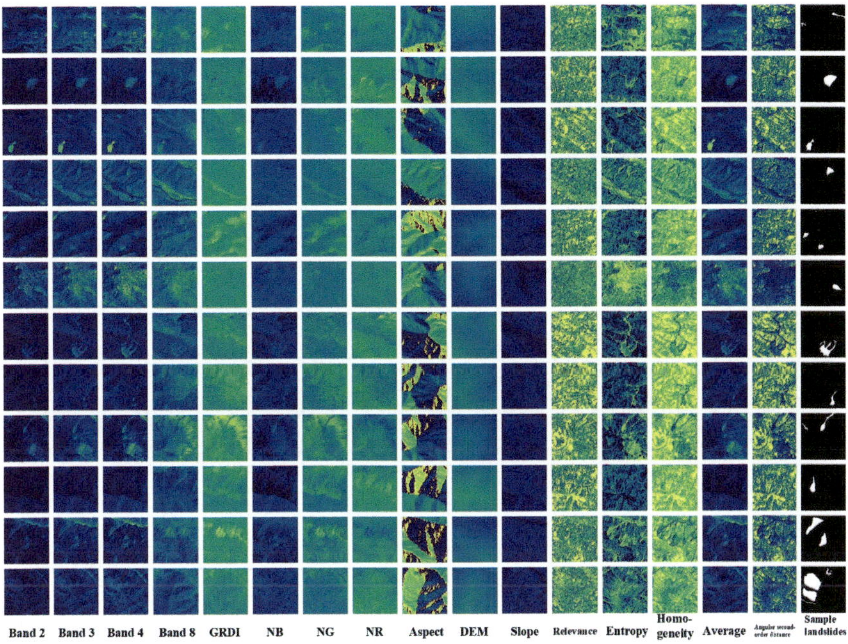

Fig. 3.23 Sample data set

3.3.2 Development of the Multi-Model Ensemble Landslide Detection Model

3.3.2.1 Backbone Architecture of the Model

To further improve segmentation accuracy, three deep learning models—FCN, U-Net, and Attention U-Net—were employed as base learners in a stacking ensemble framework, with a CNN serving as the meta-classifier (Fig. 3.24). The core idea of stacking is to integrate predictions from multiple base models using a higher-level learner, thereby reducing generalization error and improving prediction reliability. FCN is a foundational semantic segmentation model that performs pixel-wise classification using convolution, pooling, and upsampling operations, and preserves multi-scale information by element-wise addition of feature maps from different layers. U-Net enhances this by introducing skip connections that concatenate encoder and decoder features, facilitating finer localization and improved segmentation of complex structures. Attention U-Net further incorporates attention gates that adaptively emphasize relevant regions in the feature maps, allowing the model to better capture boundaries and subtle details critical for landslide detection (Oktay et al., 2018).

The outputs from the base models were concatenated and used as input features for training the CNN meta-classifier, which learned to fuse these predictions into the

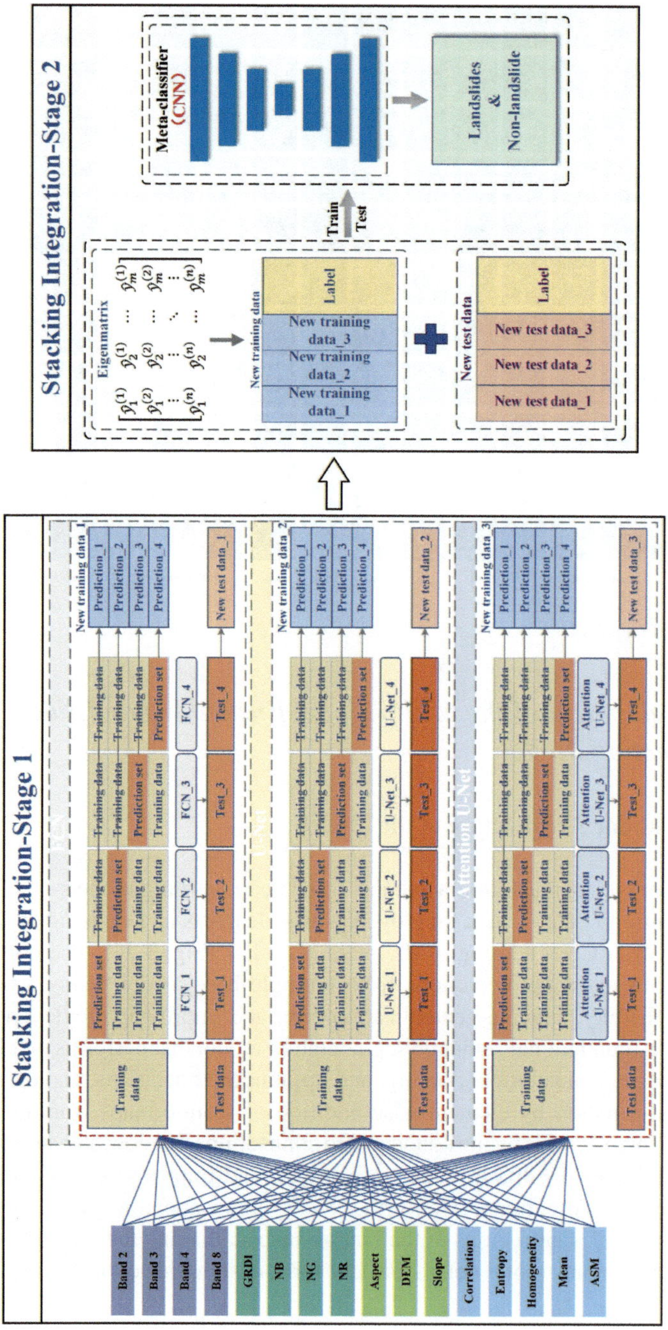

Fig. 3.24 Stacking integrated learning model structure

final landslide segmentation map. This ensemble strategy effectively leverages the complementary strengths of the individual networks, leading to more accurate and robust landslide identification.

3.3.2.2 Model Parameter Configuration and Training Procedures

To mitigate the risks of overfitting or underfitting, maximize the utilization of available data, and ensure the model's reliability and stability, K-fold cross-validation is commonly applied in training Stacking ensemble models, which also generate secondary features for training the meta-model. In this study, considering both experimental time and hardware configuration, the value of K was set to 4. During each fold, the FCN, U-Net, and Attention U-Net models were trained with an epoch set to 128, batch size of 4, learning rate of 0.00001, the Adam optimizer, and binary cross-entropy as the loss function. For training the CNN meta-classifier, the input consisted of the predictions from the base models, with real landslide samples as the ground truth labels. As the task remains binary classification, the epoch was set to 256, batch size to 8, the learning rate to 0.00001, and Adam was chosen as the optimizer with binary cross-entropy as the loss function. The hardware configuration for the experiments was as follows: Intel Core i7-10700F processor with 64 GB RAM, NVIDIA RTX 3090 GPU with 24 GB VRAM. The software configuration included CUDA version 11.1.105, Python 3.9.8, and TensorFlow 2.7.1 as the deep learning framework.

3.3.3 Landslide Detection Results

In this section, the proposed method is applied and evaluated comprehensively within the context of the Bailong river basin. Initially, models are trained using datasets comprised of various data type combinations from the training region. These trained models are then applied to landslide detection within the testing region. A comparative analysis is conducted using both visual and quantitative evaluation metrics to assess the performance of the models across different data types, thereby demonstrating the effectiveness of incorporating spectral, topographic, and texture indices. Furthermore, the effectiveness of the multi-model ensemble approach is validated by calculating and visualizing the TP, FP, and FN for both the ensemble and individual base models, allowing for a detailed comparison of the landslide detection performance from various perspectives. The results of these experiments and visual analyses provide a comprehensive evaluation of the proposed method's performance in the experimental setting, offering valuable insights for further optimization of landslide detection techniques.

3.3.3.1 Sensitivity Analysis of Spectral, Terrain, and Texture Features

To evaluate the impact of different data combinations (driving factors) on landslide detection performance, four experimental scenarios (T1, T2, T3, and T4) were defined, as outlined in Table 3.5. The models were trained using the same dataset derived from the same training region, with the testing conducted in the same region. For a holistic view of the landslide detection outcomes across different test scenarios, all sub-images of the landslide recognition results were stitched together based on geographic coordinates to form a complete map for the testing region. Additionally, four evaluation metrics—Recall, Precision, F1-score, and MIoU—were computed using the confusion matrix for quantitative mapping (Table 3.5).

Figures 3.25 and 3.26 illustrate that incorporating spectral, topographic, and texture features yielded the best recognition results, with refined detection outcomes, fewer false positives and negatives, and a more accurate representation of the actual landslides. As shown in Table 3.6, T1 exhibited the poorest performance, while T4 demonstrated the best results. As the data increased (from T1 to T4), performance metrics such as Recall, Precision, F1-score, and MIoU consistently improved, suggesting that the inclusion of spectral indices, topographic factors, and texture features positively enhances the model's performance. The Recall metric, which reflects the model's ability to identify true landslide areas, showed a steady increase as additional data was incorporated, improving recognition capability. Precision, which measures the proportion of predicted landslide areas that are truly landslides, showed some instability with increased data and could not fully capture model performance. To address this, F1-score, a comprehensive metric combining Recall and Precision, was computed. F1-score improved with increasing data, following a trend similar to Recall, and was found to increase by 2.02%, 4.14%, and 5.53% for T2, T3, and T4, respectively, compared to using only optical bands for detection. This highlights that additional data contributes to better recognition and accuracy, thereby enhancing the model's overall performance. MIoU, which measures the overlap between predicted and actual landslide areas, also showed improvement with increasing data, indicating better model accuracy with more input data.

Table 3.5 Experiment scenarios for integrated learning

	Data combinations
T1	Integrated Learning + Optical Bands
T2	Integrated Learning + Optical Bands + Spectral Indices
T3	Integrated Learning + Optical Bands + Spectral Indices + Terrain Factors
T4	Integrated Learning + Optical Bands + Spectral Indices + Terrain Factors + Texture Indices

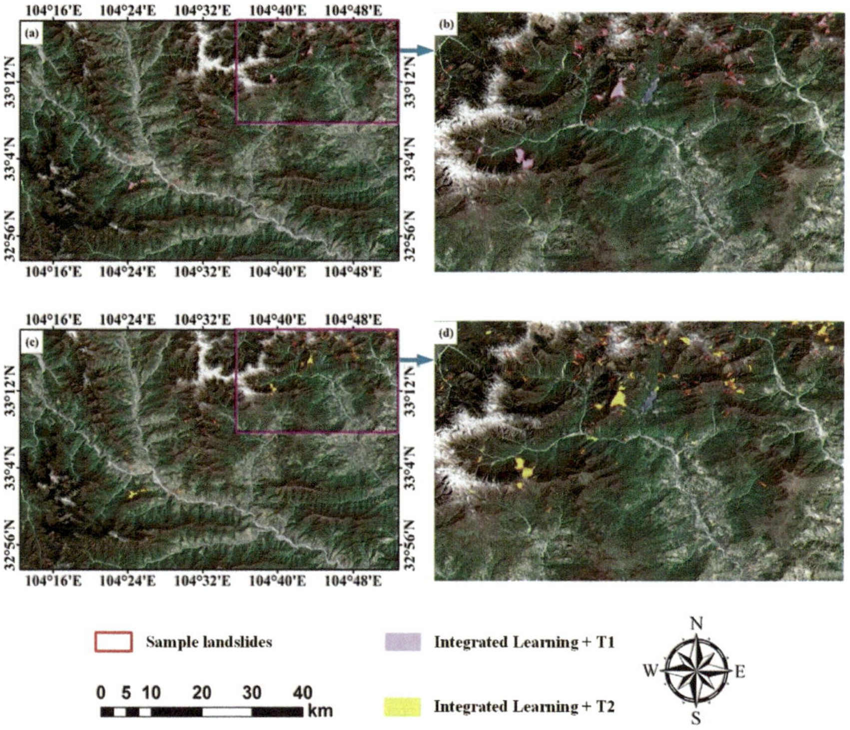

Fig. 3.25 Results of landslide identification using the integrated learning model. **a** Integrated learning + T1, **b** specific region within (**a**), **c** integrated learning + T2, **d** specific region within (**c**)

Notably, the inclusion of terrain factors led to a marked improvement in both Recall and F1-score. This enhancement is likely attributed to the complex topographic and geomorphic conditions of the Bailong River Basin, where terrain-related variables such as DEM, slope, and aspect play a critical role in the initiation and spatial distribution of landslides. The integration of terrain factors introduces supplementary information associated with surface morphology and landform characteristics, which are essential for accurate landslide detection. By incorporating terrain data, the T3 and T4 models leverage a combination of optical imagery, spectral indices, and topographic features, effectively compensating for the limitations of single-source data. This multi-source fusion enables more precise extraction of terrain characteristics and enhances the model's understanding of surface and topographic variability, thereby improving both detection capability and robustness. Consequently, the comprehensive landslide detection performance is enhanced, as reflected in the improved Recall and F1-score.

Analysis of different data combinations reveals that, constrained by the complex geographical conditions of the Bailong river basin—such as rugged terrain and dense vegetation—optical remote sensing imagery alone fails to provide sufficient landslide-related features for effective model training. In contrast, the proposed

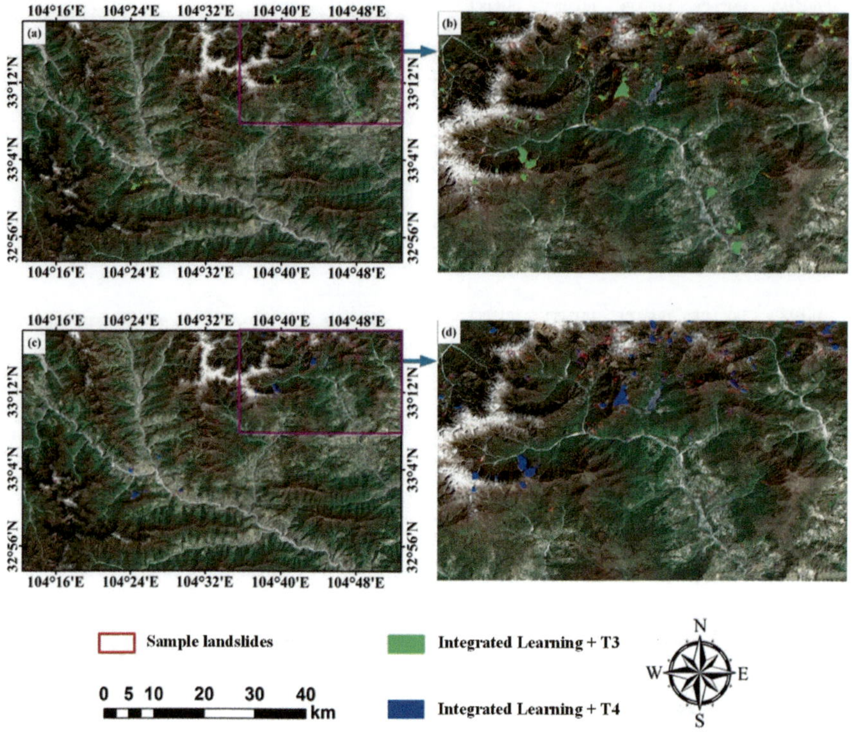

Fig. 3.26 Results of landslide identification using the integrated learning model. **a** Integrated learning + T3, **b** specific region within (**a**), **c** integrated learning + T4, **d** specific region within (**c**)

Table 3.6 Performance metrics for landslide identification results using the integrated learning model with different data combinations

Different data combinations	Recall (%)	Precision (%)	F1-score (%)	MIoU (%)
T1	54.33	62.84	58.28	70.44
T2	59.64	60.97	60.30	72.55
T3	69.17	56.87	62.42	72.55
T4	69.89	58.70	63.81	73.30

ensemble learning method based on multi-feature remote sensing data—incorporating spectral, topographic, and textural information—effectively leverages the rich and diverse characteristics of landslides from multiple data sources. This approach enables the model to learn landslide features from different perspectives, enhancing its ability to distinguish landslides from visually similar complex surface objects, and thereby improving detection accuracy. Both visualization results and quantitative evaluations confirm that the integration of additional features such as spectral indices, terrain attributes, and texture metrics significantly improves the performance of landslide detection models.

3.3.3.2 Comparative Analysis of Existing Approaches

In this study, a Stacking-based ensemble learning model was developed by integrating three deep learning architectures: FCN, U-Net, and Attention U-Net. To evaluate the model's effectiveness and demonstrate its superior performance, landslide detection results from the ensemble model were compared with those of the individual base models in the designated test area. All models successfully identified the majority of landslides. For improved clarity in visual mapping, several regions were randomly selected and magnified. The first through fifth columns of Fig. 3.27 respectively represent the detection results from the FCN, U-Net, Attention U-Net, the ensemble model, and the reference landslide samples.

Fig. 3.27, row h, the landslide detected by FCN was highly incomplete, lacking a discernible contour. Although U-Net and Attention U-Net detected most of the area, they provided suboptimal contour representation. The ensemble model, however, achieved more comprehensive detection and accurately reconstructed the landslide shape, highlighting its superior performance.

Overall, the comparison between the detection results of FCN, U-Net, Attention U-Net, and the Stacking ensemble model against the reference landslide samples demonstrates that the ensemble model effectively integrates the predictive capabilities of individual models. By leveraging their complementary strengths and mitigating their limitations, the ensemble model produces more accurate and robust landslide detection outcomes.

To further evaluate the landslide detection performance from multiple perspectives, we calculated the true positive (TP), false positive (FP), and false negative (FN) values for the FCN, U-Net, Attention U-Net, and the proposed Stacking ensemble model based on the reference landslide inventory. Representative regions were selected and visually enlarged for comparison (Fig. 3.28). As observed, the ensemble model achieves a higher number of TP pixels and fewer FP and FN pixels, indicating improved detection accuracy with reduced false alarms and missed detections.

In addition, to better illustrate the differences between predicted and actual landslides, we selected typical regions and overlaid the recognition results onto the original RGB imagery for visual interpretation (Fig. 3.29). As shown in various scenes of Fig. 3.29, the ensemble model consistently identifies more TP pixels (red areas in the fourth column, correctly predicted landslides), while reducing FP (blue-purple areas, falsely predicted landslides) and FN pixels (yellow areas, missed landslides). These results demonstrate the ensemble model's superior balance between Precision and Recall, enabling more accurate and complete identification of landslides compared to individual models. The results further highlight its effectiveness in integrating the strengths of different models and enhancing detection reliability.

Quantitative results (Fig. 3.30) further support these observations. The FCN model exhibits a low Recall (41.76%), indicating a high rate of omission, although its

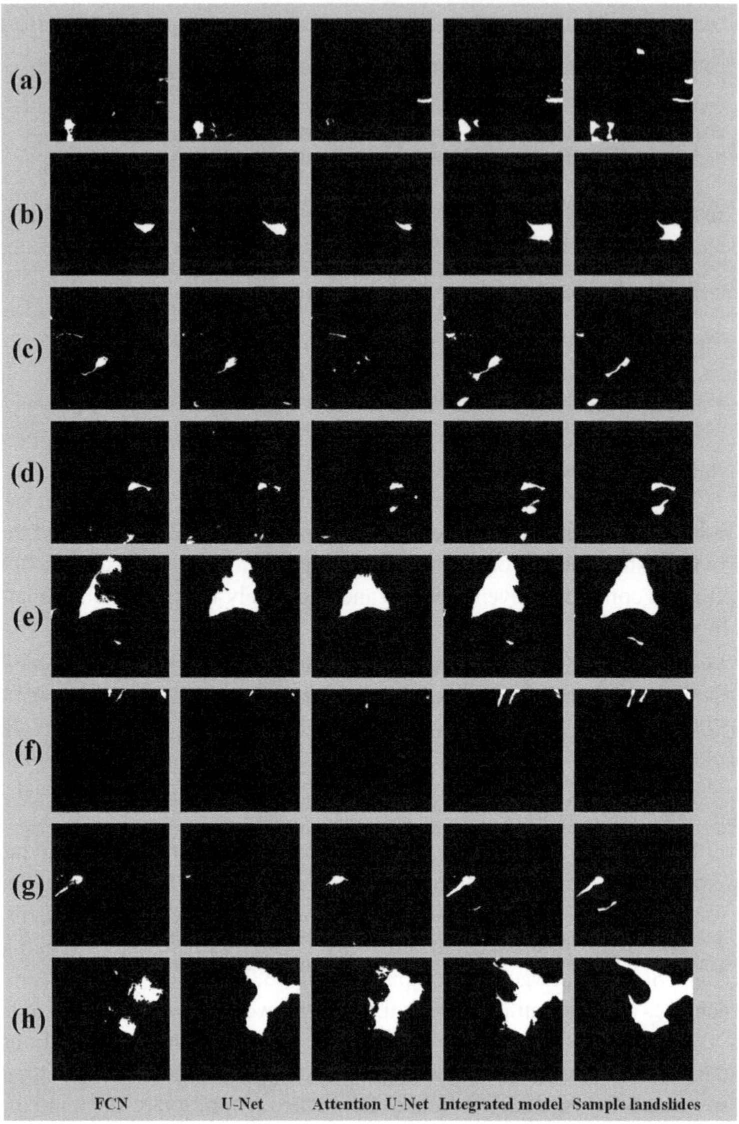

Fig. 3.27 Results of landslide identification using different models

Precision is relatively high (71.32%), suggesting fewer false alarms. The F1-score (52.60%) and MIoU (67.73%) reflect moderate overall performance. U-Net demonstrates a slight improvement in Recall (47.83%) with consistently high Precision (75.41%), resulting in better F1-score (58.83%) and MIoU (70.58%). Attention U-Net yields similar metrics to FCN, with Recall and Precision at 42.78% and 69.79%, respectively, and moderate performance in F1-score (53.05%) and MIoU (67.93%).

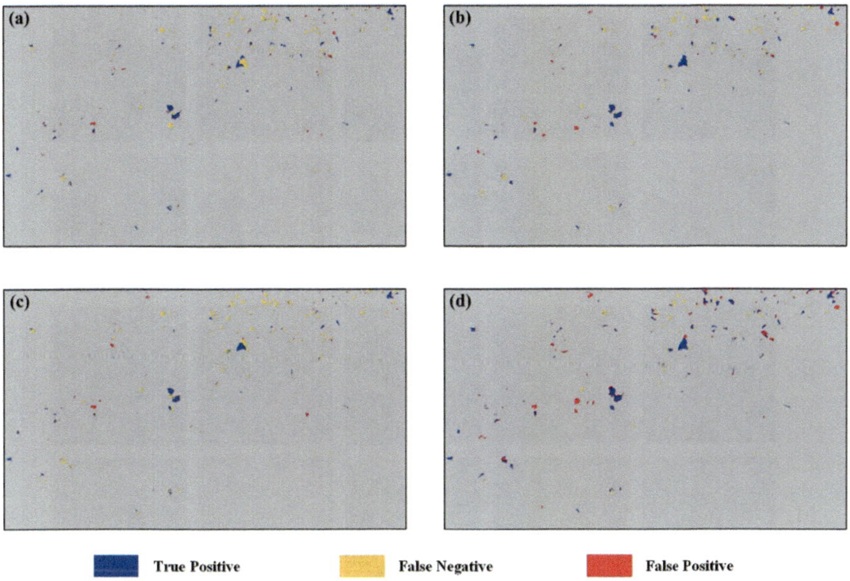

Fig. 3.28 Diagrams of TP, FN, and FP for landslide identification results of different models. **a** FCN, **b** U-Net, **c** attention U-Net, **d** integrated learning model

In contrast, the Stacking ensemble model significantly improves Recall to 69.89%, effectively reducing missed detections. Although its Precision (58.70%) is lower than that of individual models, the ensemble model still maintains a low false alarm rate. The F1-score (63.81%) and MIoU (73.30%) clearly outperform all baseline models, indicating superior landslide delineation and more robust predictive capabilities. These results confirm that the ensemble approach enhances the generalization ability and overall robustness of the landslide detection framework.

FCN, U-Net, and Attention U-Net possess distinct characteristics that make them suitable for different tasks: FCN is adept at handling features at varying spatial scales, U-Net excels in pixel-level classification, and Attention U-Net incorporates an attention mechanism that enhances the model's focus on salient feature regions. Due to their differing architectures and learning capabilities, these models are able to capture diverse features and patterns within the data. However, each individual model may perform well under certain conditions while exhibiting limitations under others, and may also be prone to specific types of errors.

To address these issues, a Stacking ensemble model was constructed, integrating multiple base classifiers (e.g., FCN, U-Net, and Attention U-Net) using a meta-classifier (CNN). This approach effectively leverages the diversity among the base models. By aggregating the outputs of multiple models, the ensemble mitigates the shortcomings of individual classifiers and reduces the risk of error propagation. The meta-classifier learns the interrelationships and weighting among the base classifiers, enabling a weighted fusion of predictions. Consequently, the ensemble model can

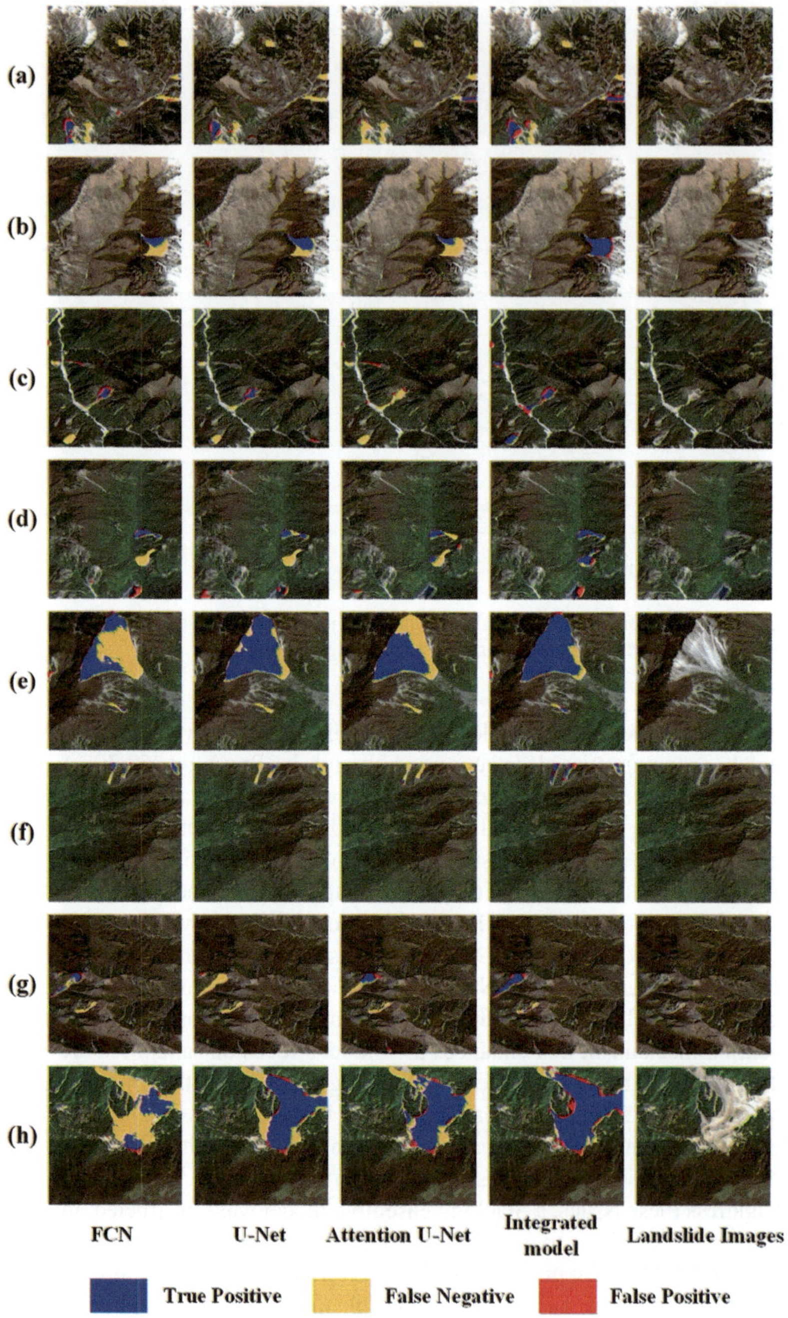

Fig. 3.29 Diagrams of TP, FN, and FP for landslide identification results in typical regions

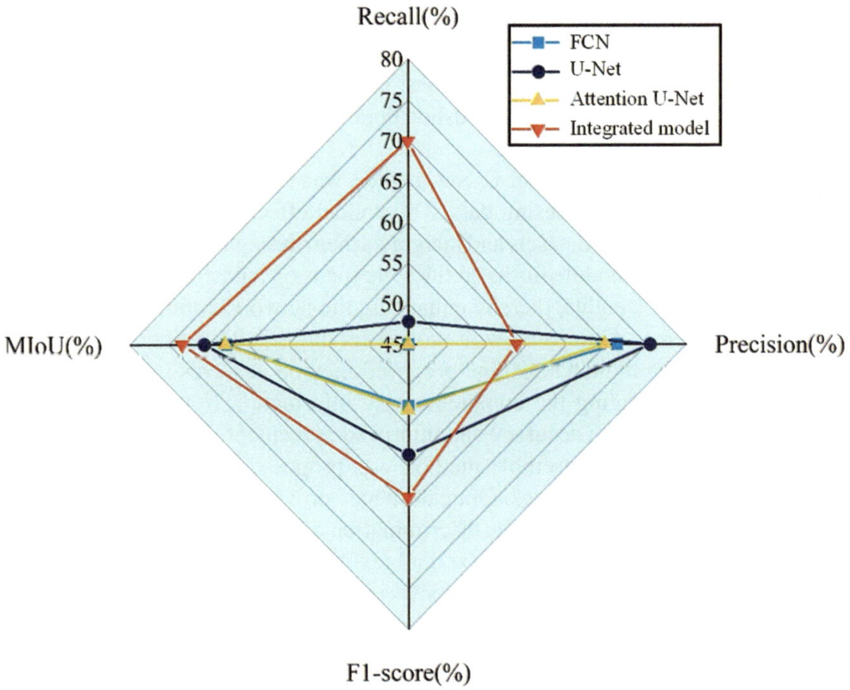

Fig. 3.30 Quantitative evaluation metrics for landslide identification results

capitalize on the advantages of each base model, enhancing generalization ability, lowering the risk of overfitting, and improving overall performance in landslide detection tasks.

3.4 Summary

This chapter presents three intelligent landslide detection approaches applied in Lanzhou and the Bailong river basin, Gansu Province, Northwest Chian: an attention-based U-Net method utilizing optical remote sensing imagery, a multi-input channel U-Net method integrating multi-source remote sensing data, and a multi-model ensemble method that incorporates spectral, topographic, and texture features. These approaches significantly enhance landslide recognition accuracy and provide reliable landslide inventories for subsequent susceptibility mapping.

First, a channel attention U-Net model based on Sentinel-2A imagery was developed for landslide detection. By integrating U-Net with SENet, the proposed model leverages the rich spectral information of Sentinel-2A and dynamically adjusts the weights of different feature channels, emphasizing those with higher relevance to landslide classification. This enables more effective feature learning. Both qualitative and quantitative evaluations demonstrate that the method better differentiates between landslide and non-landslide areas, effectively captures landslide-specific features, and accurately delineates landslide boundaries, thereby reducing both false alarms and omissions.

Second, a multi-input channel U-Net method was constructed by incorporating SAR multi-source remote sensing data. This model effectively utilizes the diverse features provided by SAR datasets and integrates them through a multi-channel architecture. The attention module further refines the focus on landslide-relevant features by assigning higher weights, thereby enhancing the network's ability to learn from complex data sources. Compared with models relying solely on optical imagery or conventional U-Net architectures, the proposed method achieves improved performance by fully leveraging the complementary information from SAR data. This results in higher detection accuracy and improved agreement with ground truth.

Lastly, a multi-model ensemble method was proposed that integrates spectral, topographic, and texture features. Optical images provide abundant spectral information, spectral indices enhance the discrimination of land cover types, topographic factors offer geomorphological context, and texture indices capture surface roughness—together contributing to comprehensive feature representation. The Stacking ensemble model combines FCN, U-Net, and Attention U-Net architectures to mitigate the limitations of individual models and to fully exploit the strengths of heterogeneous data. This approach enhances landslide feature representation and enables robust detection across complex terrain. Compared to single-model or single-source remote sensing approaches, the proposed ensemble method achieves more accurate and complete identification of landslide areas, with improved classification performance and reduced rates of false detection and omission.

References

Brauwers, G., & Frasincar, F. (2023). A general survey on attention mechanisms in deep learning. *IEEE Transactions on Knowledge and Data Engineering, 35*(4), 3279–3298. https://doi.org/10.1109/TKDE.2021.3126456

Chen, H., He, Y., Zhang, L., Yao, S., Yang, W., Fang, Y., Liu, Y., & Gao, B. (2023). A landslide extraction method of channel attention mechanism U-Net network based on Sentinel-2A remote sensing images. *International Journal of Digital Earth.* https://www.tandfonline.com/doi/abs/https://doi.org/10.1080/17538947.2023.2177359

Chen, H., He, Y., Zhang, L., Yang, W., Liu, Y., Gao, B., Zhang, Q., & Lu, J. (2024). A multi-input channel U-Net landslide detection method fusing SAR multisource remote sensing data. *IEEE Journal of Selected Topics in Applied Earth Observations and Remote Sensing, 17*, 1215–1232. https://doi.org/10.1109/JSTARS.2023.3339294

Fang, M., Zhang, J., & Xu, Z. (2011). Landslide susceptibility zoning study in lanzhou city based on gis and logistic regression model. *Remote Sensing Technology and Application, 26*(6), 845–854.

Hacıefendioğlu, K., Demir, G., & Başağa, H. B. (2021). Landslide detection using visualization techniques for deep convolutional neural network models. *Natural Hazards, 109*(1), 329–350. https://doi.org/10.1007/s11069-021-04838-y

Hu, J., Shen, L., & Sun, G. (2018). Squeeze-and-excitation networks. *IEEE/CVF Conference on Computer Vision and Pattern Recognition, 2018*, 7132–7141. https://doi.org/10.1109/CVPR.2018.00745

Kingma, D. P., & Ba, J. (2017). *Adam: A Method for Stochastic Optimization* (No. arXiv:1412.6980). https://doi.org/10.48550/arXiv.1412.6980

Lissak, C., Bartsch, A., De Michele, M., Gomez, C., Maquaire, O., Raucoules, D., & Roulland, T. (2020). Remote sensing for assessing landslides and associated hazards. *Surveys in Geophysics, 41*(6), 1391–1435. https://doi.org/10.1007/s10712-020-09609-1

Lou, P., Wu, T., Chen, J., Fu, B., Zhu, X., Chen, J., Wu, X., Yang, S., Li, R., Lin, X., Shang, C., Wen, A., Wang, D., La, Y., & Ma, X. (2023). Recognition of thaw slumps based on machine learning and UAVs: A case study in the Qilian Mountains, northeastern Qinghai-Tibet Plateau. *International Journal of Applied Earth Observation and Geoinformation, 116*, Article 103163. https://doi.org/10.1016/j.jag.2022.103163

Lyons, M. B., Keith, D. A., Phinn, S. R., Mason, T. J., & Elith, J. (2018). A comparison of resampling methods for remote sensing classification and accuracy assessment. *Remote Sensing of Environment, 208*, 145–153. https://doi.org/10.1016/j.rse.2018.02.026

Mei, H., & Zhang, J. (2010). Types and distribution of geological hazards in Lanzhou city. *South-to-North Water Transfers and Water Science & Technology, 8*(4), 53–57.

Meng, X., Chen, G., Guo, P., Xiong, N., & Janusz, W. (2013). Research of landslides and debris flows in bailong river basin: progress and prospect. *Marine Geology & Quaternary Geology, 33*(4), 1–15.

Metternicht, G., Hurni, L., & Gogu, R. (2005). Remote sensing of landslides: An analysis of the potential contribution to geo-spatial systems for hazard assessment in mountainous environments. *Remote Sensing of Environment, 98*(2), 284–303. https://doi.org/10.1016/j.rse.2005.08.004

Niu, Z., Zhong, G., & Yu, H. (2021). A review on the attention mechanism of deep learning. *Neurocomputing, 452*, 48–62. https://doi.org/10.1016/j.neucom.2021.03.091

Oktay, O., Schlemper, J., Folgoc, L. L., Lee, M., Heinrich, M., Misawa, K., Mori, K., McDonagh, S., Hammerla, N. Y., Kainz, B., Glocker, B., & Rueckert, D. (2018). *Attention U-Net: Learning Where to Look for the Pancreas* (No. arXiv:1804.03999). https://doi.org/10.48550/arXiv.1804.03999

Panwar, H., Gupta, P. K., Siddiqui, M. K., Morales-Menendez, R., Bhardwaj, P., & Singh, V. (2020). A deep learning and grad-CAM based color visualization approach for fast detection of COVID-19 cases using chest X-ray and CT-Scan images. *Chaos, Solitons & Fractals, 140*, Article 110190. https://doi.org/10.1016/j.chaos.2020.110190

Selvaraju, R. R., Cogswell, M., Das, A., Vedantam, R., Parikh, D., & Batra, D. (2017). Grad-CAM: visual explanations from deep networks via gradient-based localization. *IEEE International Conference on Computer Vision (ICCV), 2017*, 618–626. https://doi.org/10.1109/ICCV.2017.74

Singaravel, S., Suykens, J., & Geyer, P. (2018). Deep-learning neural-network architectures and methods: Using component-based models in building-design energy prediction. *Advanced Engineering Informatics, 38*, 81–90. https://doi.org/10.1016/j.aei.2018.06.004

Wang, X., Hu, D., Tang, H., & Shu, Y. (2014). Extraction of landslide information from airborne polarimetric SAR images based on Bayes decision theory. *Remote Sensing for Natural Resources, 26*(2), 121–127.

Yang, X. (2020). An overview of the attention mechanisms in computer vision. *Journal of Physics: Conference Series, 1693*(1), Article 012173. https://doi.org/10.1088/1742-6596/1693/1/012173

Yi, Y., & Zhang, W. (2020). A new deep-learning-based approach for earthquake-triggered landslide detection from single-temporal RapidEye satellite imagery. *IEEE Journal of Selected Topics in Applied Earth Observations and Remote Sensing, 13*, 6166–6176. https://doi.org/10.1109/JSTARS.2020.3028855

Yoshimatsu, H., & Abe, S. (2006). A review of landslide hazards in Japan and assessment of their susceptibility using an analytical hierarchic process (AHP) method. *Landslides, 3*(2), 149–158. https://doi.org/10.1007/s10346-005-0031-y

Yu, X., Wu, X., Luo, C., & Ren, P. (2017). Deep learning in remote sensing scene classification: A data augmentation enhanced convolutional neural network framework. *Giscience & Remote Sensing, 54*(5), 741–758. https://doi.org/10.1080/15481603.2017.1323377

Zhang Y. (2018). *Detecting Ground Deformation and Investigating Landslides Using InSAR Technique—Taking Middle Reach of Bailong River Basin as an Example.* Lanzhou University.

Zhang, Y., Meng, X., Chen, G., Qiao, L., Zeng, R., & Chang, J. (2016). Detection of geohazards in the Bailong River Basin using synthetic aperture radar interferometry. *Landslides, 13*(5), 1273–1284. https://doi.org/10.1007/s10346-015-0660-8

Chapter 4
A Graph Neural Network Landslide Susceptibility Mapping Method Constrained by Geographical Environment Consistency

Facing the fact that the existing data-driven machine learning landslide susceptibility mapping methods fail to fully express the spatio-temporal correlation between landslides and disaster-prone environments in complex geological environments, the problems of sample imbalance and insufficient feature expression remain prominent. Moreover, the existing landslide susceptibility mapping models have a single learning mode and insufficient feature learning ability, and are unable to deeply explore the potential associated features of landslides. It is difficult to adapt to the complex heterogeneous geographical environment. This study focuses on the mapping of landslide susceptibility in graph neural networks under the constraint of geographical environment consistency. By constructing geographical nodes and weighted graph models, the precise mapping of landslide susceptibility has been achieved. Finally, a graph neural network model integrating the attention mechanism was further proposed. By strengthening the dynamic attention to the importance of environmental features, the accuracy of wide-area landslide susceptibility mapping was improved.

4.1 Research Data

This study takes the Bailong River Basin in the southeast of Gansu Province, Northwest China as an example. The topography and landforms in this basin are complex, with alternating distribution of alpine canyon landforms and loess landforms. The netectonic movement is intense, active faults are developed, and weak rock and soil masses (such as calenite) are widely distributed, forming a fragile and sensitive geological environment. At the same time, seismic activities are frequent, and the river valleys are deep. It provides favorable conditions for the occurrence of landslide disasters, resulting in frequent occurrences of such disasters. The overview of the study area is as described in Sect. 3.1.1.1.

Y. He, *InSAR and Deep Learning in Landslides Research: Intelligent Identification, Risk Assessment and Susceptibility Mapping*, https://doi.org/10.1007/978-981-96-9132-6_4

Table 4.1 Data sources of Sentinel-1A

Orbit number	Direction	Time	Angle	Azimuth angle	Resolution
Path: 55 Frame: 102, 107	Ascending	2019.10.7–2023.11.27	36.79°	−12.95°	5 m × 20 m
Path: 62 Frame: 479, 484	Descending	2019.11.12–2023.11.27	39.72°	192°	5 m × 20 m

4.1.1 *Sentinel-1A*

The Sentinel-1 satellite system, as the core Earth observation platform of the European Space Agency (ESA)'s Gopernicus program, consists of the Sentinel-1A and 1B binary satellites (with an orbital phase difference of 180°). All-weather and all-time monitoring is achieved through C-band Synthetic Aperture Radar (SAR). This satellite system, relying on high-precision radar interferometric measurement technology, is widely applied in fields such as Marine environment, soil protection and surface deformation inversion. This paper adopts the Interferometric Wide swath (IW) data of the Sentinel-1A satellite. This mode is characterized by a width of 250 km, a resolution of 5 m (range direction) × 20 m (azimuth direction), and a dual-polarization combination of VV + VH. Combined with the incident Angle range of 19–47°, millimeter-level surface deformation information can be obtained by penetrating clouds and rain (Hakim et al., 2023). Research, in October 2019 to November 2023 cover the target area of 432 single look complex images of ascending and descending tracks (https://scihub.copernicus.eu/, accessed on May 6, 2025), through a long phase analysis supported the geological disaster monitoring (Table 4.1).

4.1.2 *Landslide Database*

This paper uses various sources such as remote sensing images, field investigations and data from the China Geological Survey to compile historical landslide data in the study area from 2018 to 2022. The enumerated landslides are mainly soil-based landslides, with the landslide area ranging from 824 to 4.4 × 106 m^2. These landslides are mainly triggered by seasonal rainfall and seismic activities (Bai et al., 2012), and are distributed on both sides of rivers and valleys.

4.1.3 Landslide Influencing Factors

The genesis mechanism of landslides in the Bailong River Basin is complex. Based on the existing achievement data and previous research results (Bai et al., 2012), this paper selects a total of 12 landslide influencing factors from four aspects: topography, geology, hydrology and human engineering activities. The spatial distribution of the landslide influencing factors is shown in Table 4.2. Among them, for distance-related factors, such as faults, rivers and roads, the influencing factors are obtained through the calculation of Euclidean distance (Zhang et al., 2024a). Furthermore, due to the different resolutions of different data, the scales of the extracted landslide factors are different. To facilitate the processing of subsequent data, the ArcGIS software is used to uniformly resampled to a resolution of 30 m × 30 m. The specific data sources are shown in Table 4.2 as follows.

The topographic factors include Elevation, slope and aspect, among which the slope and aspect are calculated by the DEM. The surface of the Bailong river basin is highly undulating, the terrain is perilous and the mountainous features are prominent. Elevation and slope can provide certain potential energy for the occurrence of landslides. Landslides are more likely to occur in areas with higher elevation and steep mountain slopes (Guo et al., 2015). The aspect determines the intensity of solar radiation received by the slope surface. The differences in temperature conditions between the positive and negative slopes can lead to differences in the water content, weathering degree and physical and mechanical properties of the slope soil, which have a certain impact on the occurrence of landslides (Sameen et al., 2020).

Table 4.2 Data sources of landslide influencing factors, reprinted from Zhang et al. (2024a), copyright 2025, with permission from Elsevier

Data	Factors	Resolution	Source
Topographical	Elevation	12.5 m	https://vertex.daac.asf.alaska.edu accessed on May 6, 2025
	Slope		
	Aspect		
Geological	Lithology	1:1000,000	http://geodb.cgs.gov.cn/ accessed on May 6, 2025
	Soil type	1000 m	http://www.fao.org/soils-portal accessed on May 6, 2025
	Distance from faults	30 m	
Hydrological	Annual average precipitation	1000 m	http://data.tpdc.ac.cn accessed on May 6, 2025
	TWI	12.5 m	
	Distance from rivers	30 m	
Human activity	Land use types	30 m	http://www.geodata.cn/ http://globallandcover.com/ accessed on May 6, 2025
	NDVI	1000 m	
	Distance from roads	30 m	

Geological factors affect the occurrence of landslides by altering the strength and permeability of rock masses and soil (Xiao et al., 2018). The Bailong river basin is traversed by the West Qinling Fault Zone, and due to the superposition of multiple tectonic movements, complex large-scale folds and faults have been formed. The closer to the fault, the more vulnerable the area is to earthquakes and fault activities. Lithology interferes with the gestation of landslides by directly influencing the mechanical properties of bedrock and deposits. The texture, structure and moisture content of different soil types vary, resulting in differences in soil shear resistance capacity, which has a certain impact on the development of landslides (Huang et al., 2019) (Fig. 4.1).

Human engineering activity factors are subjective factors influencing the occurrence of landslides. Land use and road construction reflect human beings' utilization and alteration of natural conditions. Large-scale reclamation and construction have changed the natural form of the earth's surface, accelerated soil erosion and soil erosion, and increased the risk of landslides. The normalized difference vegetation index (NDVI) reflects the growth and coverage degree of vegetation in the area. Areas with high coverage usually have a lower possibility of landslides (Yunus et al., 2020).

Whether the landslide influencing factors are independent of each other determines whether the trained model has sufficient robustness and stability. Therefore, a mathematical analysis of the collinearity problem among landslide factors is required. In this paper, the variance expansion factor (VIF) and Tolerance (TOL) are adopted to test the degree of collinearity among landslide factors (O'brien, 2007). The formula is as follows:

$$VIF = \frac{1}{1 - A^2} = \frac{1}{T} \tag{4.1}$$

Here, A^2 represents the variance among the variables, and VIF represents the ratio of the variance of the estimated regression coefficients when the variance of the regression coefficients estimated by the least square method does not have a linear relationship with the assumed independent variables. As the standard deviation increases, the degree of collinearity also increases accordingly. When $VIF < 10$ or $T > 0.1$, it indicates that the selected independent variable has good independence.

The analysis results are shown in Fig. 4.2. The VIF values of the 12 landslide influencing factors selected in this paper are all between 1 and 2. The maximum value occurs at the elevation (1.78), and the minimum value occurs at the land use type (1.04). The tolerance T value is between 0.5 and 1, with the maximum value being 0.96 and the minimum value being 0.56. The results show that the variance expansion coefficients and tolerances of the selected 12 influencing factors all meet the critical value requirements, indicating that they have good independence and there is no collinearity problem. To sum up, the 12 landslide influencing factors selected in the study area of this paper can all be substituted into the subsequent model learning and result evaluation, which can ensure the validity and accuracy of the data.

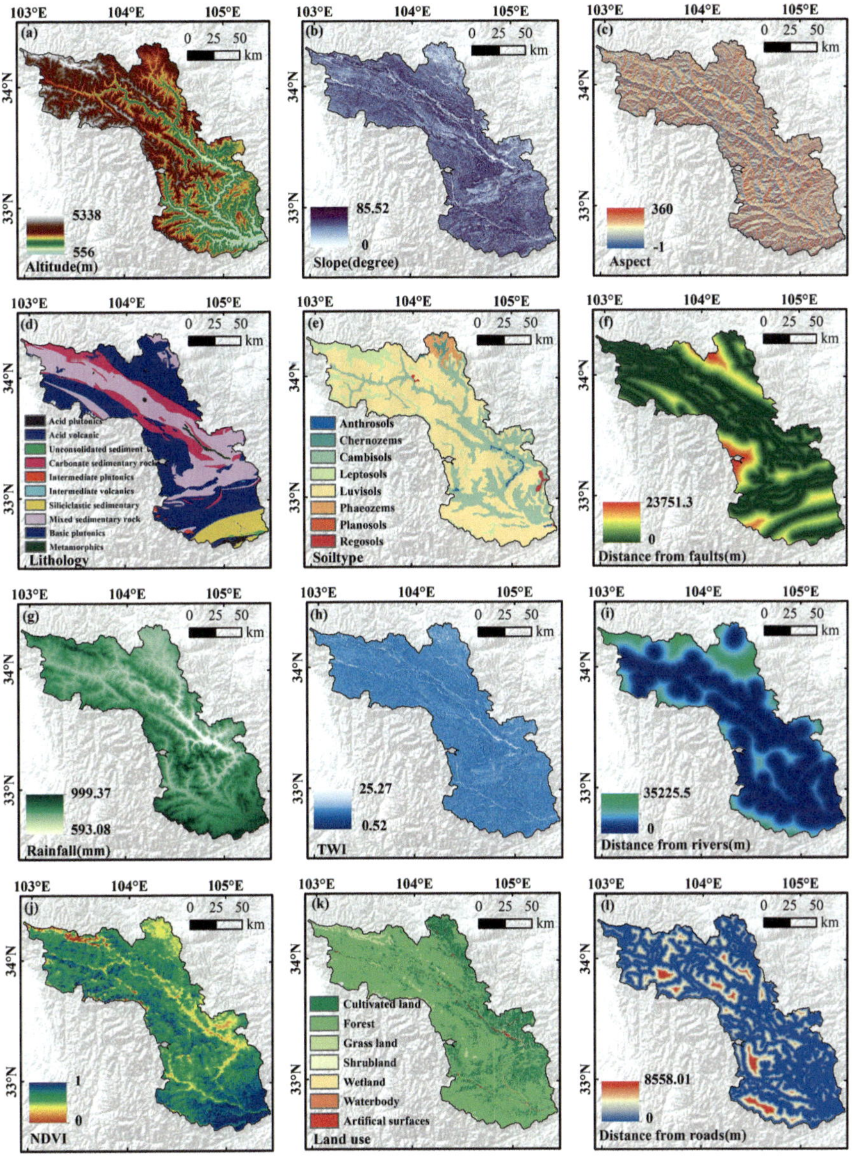

Fig. 4.1 Landslide influencing factors: **a** elevation, **b** slope, **c** aspect, **d** lithology, **e** soil type, **f** distance from faults, **g** annual average precipitation, **h** topographic wetness index, **i** distance from rivers, **j** normalized difference vegetation index, **k** land use types, **l** distance from roads, reprinted from Zhang et al. (2024a), copyright 2025, with permission from Elsevier

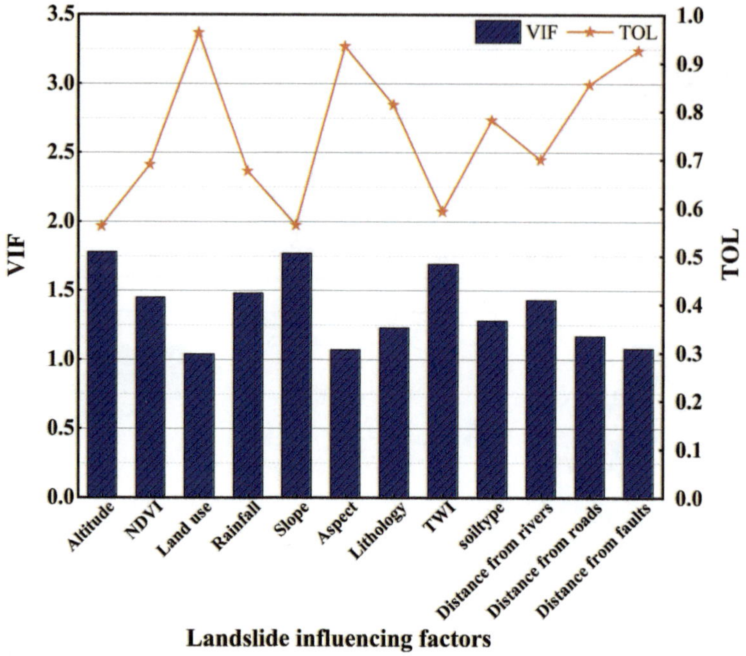

Fig. 4.2 Results of the multicollinearity analysis of LIFs, reprinted from Zhang et al. (2024a), with permission from Elsevier

4.1.4 Slope Unit

In the mapping of landslide susceptibility, constructing a scientific and reasonable evaluation unit is a crucial basic task. This process aims to standardize and integrate the data of multi-source and heterogeneous landslide disaster factors for analysis. Specifically, due to the significant differences among various slide-related factors in terms of attribute characteristics, spatial resolution, expression scale, and minimum mapping units, how to effectively integrate these complex and diverse data has become the core challenge of the assessment work. The key to constructing the evaluation unit lies in converting the factor data from different sources to a consistent spatial framework through a unified spatial division strategy. This process not only requires the classification and simplification of similar areas to reduce data redundancy, but also must ensure that while reducing data complexity, the most critical characteristic information affecting the occurrence of landslides is retained. The ideal design of the evaluation unit should pursue the maximization of internal homogeneity and the maximization of differences among units, thereby achieving a balance between data simplification and the retention of key information. Therefore, determining a reasonable and unified mapping unit is an important basis for the mapping of landslide susceptibility. The mainstream landslide mapping units include grid units and slope units.

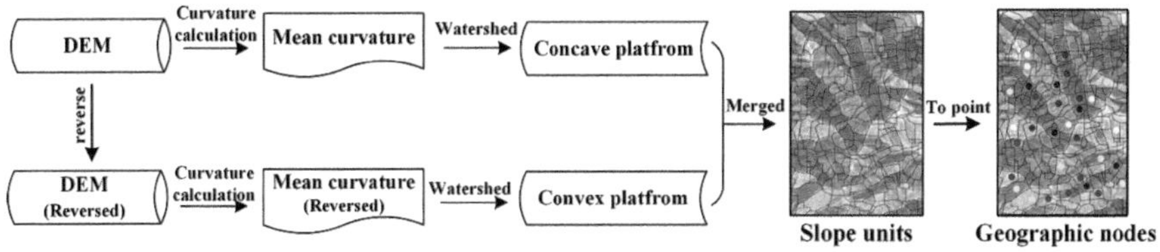

Fig. 4.3 Procedure of geographic node construction, reprinted from Zhang et al. (2024a), copyright 2025, with permission from Elsevier

The grid unit structure is simple and convenient for calculation, but it is difficult to precisely reflect the geomorphic morphology and the characteristics of the landslide disaster-prone environment, and thus cannot be applied in large-scale studies. The Bailong river basin has complex terrain, vast area and large surface undulations. Using grid units to construct geographical nodes cannot accurately summarize the geomorphic characteristics and geological environment features of the landslide. Slope units are topographic units defined by ridge lines and valley lines, which are related to the hydrological conditions and processes that shape the topography and landforms, and can accurately reflect the topographic and landform characteristics of different slope sections (Romstad & Etzelmüller, 2012). At present, there are two methods for dividing slope units: hydrological analysis method and the curvature watershed method. During the division process of hydrological analysis, a large number of fine broken surfaces and unreasonable long strip-shaped surfaces will occur, which requires a lot of manual correction work in the later stage. The process is cumbersome and the efficiency is extremely low. The curvature watershed method makes up for the drawback that the hydrological analysis method cannot delineate slopes consistent with the landform in the flat area. Meanwhile, the created slope units have uniform slope and aspect, which can meet the basic assumption of uniformity in landslide mechanical analysis. Therefore, in this paper, the curvature watershed method is adopted to construct slope units. A total of 2.2×10^5 slope units are generated, and the area of each slope units is between 1,000 m^2 and 6.4×10^5 $m^{2.}$ The construction process of the slope units is shown in Fig. 4.3.

4.2 Construction of Geographical Nodes for Landslide-Prone Environment Constrained by Terrain

4.2.1 Construction of Geographical Nodes

The landslide scene is complex and the landslide-prone environment is heterogeneous. An effective data simplification scheme is needed to express the landslide-prone environment without losing the characteristic information. This paper proposes to utilize geographical nodes to maximize the uniformity of landslide characteristics and the heterogeneity between units, providing support for the subsequent construction of maps constrained by environmental consistency and the mapping of landslide susceptibility. The slope unit constructed in Sect. 4.1.4 was adopted as the evaluation unit to generate geographical nodes, and a total of 2.2×10^5 geographical nodes were obtained. Meanwhile, the creation process of slope units gathers areas with similar slopes and aspects, thereby preserving important topographic undulation features.

4.2.2 Geographical Node Feature Extraction

The attributes of slope units are important characteristics of graph nodes. This paper is based on the ArcGIS platform and adopts the method of partitioned statistics to calculate the attribute characteristics of each node. The influencing factors of landslides generally have two types: continuous and discrete, and two different statistical methods need to be adopted for statistics. The 12 landslide influencing factors selected by us in the Bailong river basin were classified into 8 factors, namely elevation, slope, distance from the river, distance from the fault, distance from the road, topographic wetness index, normalized difference vegetation index and annual average precipitation, as continuous factors, and 4 factors, namely aspect, lithology, soil type and land use type, as discrete factors. For the continuity factor, calculate the average value of the influencing factor within the slope unit (Canavesi et al., 2020); For discrete factors, they are first reclassified according to the category, and then the category with the highest occurrence frequency of the original influencing factor within the slope unit unit is counted. Finally, the attribute characteristics of the geographical nodes of the landslide influencing factors in the Bailong River Basin were extracted. The specific processing flow is shown in Fig. 4.4.

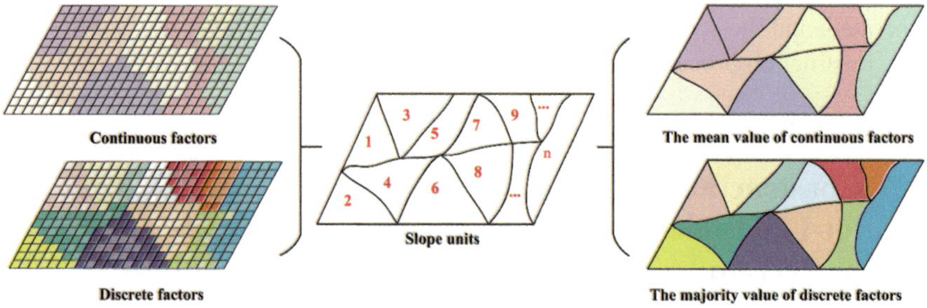

Fig. 4.4 Attributed feature extraction based on slope units, reprinted from Zhang et al. (2024a), copyright 2025, with permission from Elsevier

4.3 Construction of a Directed Weighted Graph of Node Features Constrained by Geographical Environment Consistency

4.3.1 Graph Edge Reconstruction Constrained by Geographical Environment Consistency

The graph structure E in $G = (V, E)$ is adopted to express the relationship between node V, convey the landslide information in different geographical environments, and establishing an appropriate relationship is the key to the construction of the landslide susceptibility mapping graph. The relationships of geographical units mainly include spatial relationships, feature similarity relationships and interaction relationships. Graphs based on spatial relationships can be directly generated through the topological relationships of slope units, with a fixed neighborhood size, which is convenient for calculation. However, the landslide characteristics of adjacent elements are not necessarily related. The graph structure based on spatial relationships can only play a role in smoothing the features in space. Graphs based on feature relationships can cluster nodes with similar features. Appropriate feature relationships are the key to implementing such graphs. Graphs based on interaction relationships can express the landslide mechanism among geographical units and have strong interpretability. However, due to the overly complex landslide mechanism, its interaction relationship is difficult to describe accurately. Therefore, this paper selects the feature relationship to construct the graph.

For the mapping of landslide susceptibility, the relationships among geographical nodes represent their similarity to the characteristics of landslides. Based on this idea, this paper designs environmental rules to explicitly describe the relationship E between geographical nodes V. To ensure the effective association among geographical nodes, it is necessary to construct a reasonable graph $G = (V, E)$. In this paper, environmental rules are designed to display and describe the relationship E between geographical nodes V, mainly from two aspects: the similarity of environmental factors and the similarity constraints of node features to constrain the construction of the graph. The specific construction process is shown in Fig. 4.5.

(1) Environmental factor similarity constraint
The topographic and geomorphic conditions where the landslide is gestation are the basis for its gestation. Unstable environmental conditions are the basic conditions for the occurrence of landslides, which determine the basic probability of landslide formation. That is, a consistent environment is the basic condition for node connection. From the perspective of the slope system, as a form of slope damage, the carrier of a landslide is composed of substances that remain relatively stable over a long period of time and change very little. Within the same landslide, they have consistent lithology and soil types, and these substances directly determine or affect the long-term geological conditions of the landslide (Schlögel et al., 2018). In addition, the geological environment conditions of

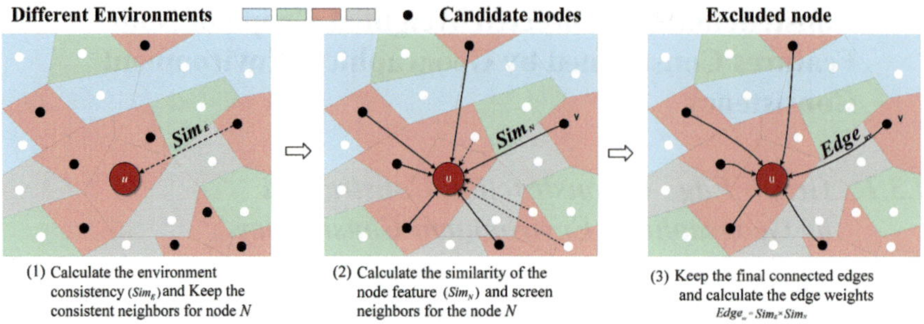

Fig. 4.5 The process of graph construction, reprinted from Zhang et al. (2024a), copyright 2025, with permission from Elsevier

landslides are also affected by the erosion of faults and rivers, and the historical landslides in the study area show the characteristics of distribution along faults and rivers. Therefore, in this paper, lithology, soil types, distance from faults and distance from rivers are taken as factors of environmental constraints to achieve the division of unstable environments. To measure the consistency of the environments where nodes u and v are located, cosine similarity (Eq. (4.2)) is adopted to calculate the similarity $Sim_E^u(v)$ of the characteristic vectors of the environmental factors of nodes x_u and x_v. Each geographic node u is only allowed to connect to the node with the most consistent environmental category. That is, when $Sim_E^u(v)$ is equal to MAX($Sim_E^u(v)$, $v \in V$), v is added to the candidate set $N(u)$ of the connected nodes of u.

$$\cos\theta_{uv} = \frac{\overrightarrow{x_u} \cdot \overrightarrow{x_v}}{|x_u| \cdot |x_v|}, \forall u, v \in V \tag{4.2}$$

$$Sim_E^u(v) = 1 - \cos\theta_{\text{uv}} \tag{4.3}$$

(2) Node feature similarity constraint
After the environmental consistency constraint, for each node u, there are still a large number of candidate nodes in $N(u)$. Nodes with different characteristics are meaningless for the overall landslide feature. Therefore, it is necessary to further calculate the similarity of node features to filter the candidate nodes. In this paper, elevation, slope, aspect, terrain wetness index, annual average cumulative rainfall, normalized difference vegetation index, distance from the road and land use type are taken as node features. Cosine similarity (Eq. (4.4)) is adopted to calculate the similarity of the sum of geographical node feature vectors $x_{u'}$ and $x_{v'}$ of nodes u and v to filter out redundant nodes. For each node, Only the candidate node with the highest similarity is retained. That is, when $Sim_N^u(v)$ is equal to $MAX(Sim_N^u(v), \forall v \in N(u))$, v can be added to the final neighbor $N'(u)$ until the size of $N'(u)$ reaches the η threshold, and finally v is

generated, $\forall v \in N'(u)$, to the edge of u. In this article, setting $\eta = 10$ means that a node has 10 neighbors. To consider the inherent characteristics of the nodes in the calculation, a self-loop is added to each node, namely.

$$\cos\theta_{u'v'} = \frac{\overrightarrow{x_{u'}} \cdot \overrightarrow{x_{v'}}}{|x_{u'}| \cdot |x_{v'}|}, \quad \forall u, v \in V \tag{4.4}$$

$$Sim_N^u(v) = 1 - \cos\theta_{u'v'} \tag{4.5}$$

4.3.2 Construction of Edge Attributes for Directed Weighted Graphs

The original model of the graph neural network performs equivalent processing on the feature information of each neighbor in the graph. However, in different disaster-prone environments, the influence of each neighbor node on the central node is different. Equivalent processing will cause the model to fail to learn the most effective features (Zeng et al., 2022). To this end, this paper adopts the method of adding weights to each neighbor during the process of aggregating neighbor information in the model to help the model better understand the relationships of nodes in the graph. The product of the similarity of environmental factors and the similarity of node features between the central node and its neighboring nodes is taken as the weight of the edge to comprehensively reflect the degree of influence of neighboring nodes on the central node. The calculation formula is as follows.

$$Sim_E^u(v) = 1 - \cos\theta_{uv} \tag{4.6}$$

4.3.3 Sampling Strategy Design Based on Environmental Similarity Constraints

The quality of the training sample set has a crucial influence on the accuracy of the landslide susceptibility mapping (Fu et al., 2022). The training samples are concentrated. The positive samples are based on the historical landslide cataloging and sampled in the landslide occurrence area. Their quality is relatively high and can fully reflect the typical geographical environmental conditions of landslide occurrence. However, negative samples cannot be obtained directly. Usually, an equal number of samples as the positive samples are randomly selected in the study area. The collected negative samples are not representative, have no typical significance, and are highly likely to contain potential landslide points among them. These false negative samples will reduce the quality of the training samples, thereby affecting the accuracy of the landslide susceptibility mapping. Therefore, in order to improve the reliability of the landslide susceptibility mapping, this paper selects non-landslide samples based on the similarity constraint environment divided in the previous text. The training dataset should not only ensure the balance of the number of positive and negative samples of landslides, but also maintain a balance on different landslide features to ensure that the model can learn more accurate, reliable and rich features. The specific steps are as follows: Based on the 175 types of environments classified above, use the cataloged landslide data as positive samples. In each environment, select negative samples in proportion to the positive samples respectively to ensure that the obtained dataset contains landslide and non-landslide samples of all environmental types, and obtain the sample dataset.

4.4 Construction of GraphSAGE Landslide Susceptibility Mapping Model Based on Environmental Information Aggregation

4.4.1 Overview of Graph Neural Networks

Graph neural networks refer to the general term for algorithms that use neural networks to learn graph-structured data, extract and mine features and patterns in graph-structured data, and meet the requirements of graph learning tasks such as clustering, classification, prediction, segmentation and generation. Graph neural networks can be divided into frequency-domain graph convolutional neural networks based on graphs and spatial-domain graph convolutional neural networks based on Spaces. Bruna et al. (2014) proposed the graph convolutional neural network based on the graph in 2014. Combining the graph theory and starting from the convolution theorem, graph convolution was defined in the frequency domain. The specific approach is to first convert the signal from the original space to the frequency domain by using the Fourier transform, then perform multiplication on the signal in the frequency domain, and finally convert the signal back to the original space by using the inverse Fourier transform, thereby achieving graph convolution. The spatial-based graph convolutional neural network was first proposed by Micheli (2009). By combining non-recursive layers, it solves the interdependence of graphs and inherits the idea of message passing at the same time. The graph convolution operation of the spatial method is similar to the traditional convolution operation, and the graph convolution is defined based on the spatial relationship of nodes. Specifically, starting from the node set, define the aggregation operation to aggregate the information of each node and its neighboring nodes; define the aggregation operation to merge the aggregated information groups and update the information of the nodes (Fig. 4.6).

Frequency-domain graph convolution requires converting data from the spatial domain to the frequency domain for processing. It demands that the graph structure remain unchanged and cannot handle directed graphs. In contrast, spatial domain

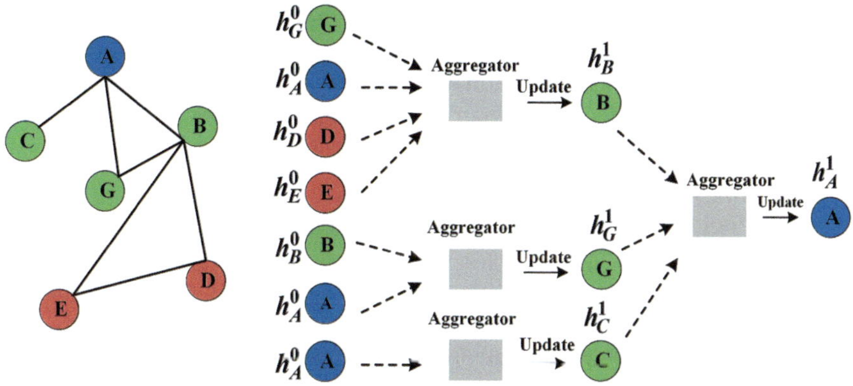

Fig. 4.6 The principle of GNN based on spatial domain, reprinted from Zhang et al. (2024a), copyright 2025, with permission from Elsevier

graph convolution does not rely on graph theory and convolution theorems. It directly defines the convolution operation in space, avoiding the data transformation process. Therefore, we adopt the spatial-based graph convolutional neural network to achieve the mapping of landslide susceptibility. At present, the mainstream Graph neural networks include graph conventional networks (GCN), Graph Sample and aggregate (GraphSAGE), and Graph attention network (GAT). Among them, GraphSAGE is a graph-based inductive learning framework. By using the node feature information and structural information, sampling and aggregating the features of neighboring nodes and vertices from the local neighbors of the vertices (Zeng et al., 2022), the Graph Embedding of the vertices is obtained. The GraphSAGE model has the following characteristics: It can be distributed trained on large-scale graphs; there is no need to consider the information of the entire graph. For new nodes, the previously trained model can be directly applied to calculate the embedding features of the nodes without retraining the changed graph. These characteristics are highly advantageous in the mapping of landslide susceptibility in large-scale areas. Moreover, in the mapping of newly added areas in a small range and the re-evaluation after the update of local node labels, reliable results can be directly generated without repeated calculations. Therefore, this paper introduces the GraphSAGE model to generate node embeddings, predict node labels, and complete the mapping of landslide susceptibility.

4.4.2 Construction of GraphSAGE Model for Aggregating Environmental Information

The GraphSAGE model, as a representative of the spatial domain, is a typical inductive learning framework. By extracting the feature information of graph neighborhood nodes through neighbor sampling and information aggregation, it makes the representation learning of nodes on large graphs possible and is widely used in various node classification tasks. Therefore, in this paper, the GraphSAGE model is adopted to aggregate feature information from the local domain of nodes, generate node embeddings, and predict the landslide susceptibility of each node.

(1) GraphSAGE
The core idea of the GraphSAGE model is to generate the embedding vector of the target node by learning a function that aggregates the representation of neighboring nodes. The information aggregated on the nodes is utilized to achieve the purpose of predicting the node labels, as shown in Fig. 4.7.

First, perform uniform sampling on node u with a fixed neighborhood size $N(u)$; Then, the feature information of the neighbor nodes is fused through the aggregation function, that is $h_{N(u)}^{l+1}$, the representation of the neighbor node $N(u)$ of node u in the $l+1$ layer (Eq. (4.7)). Finally, it is concatenated with the current representation h_u^l of the node through the concat operator, multiplied by the weight matrix W, and then fed to the fully connected layer using the nonlinear activation function σ to generate the representation h_u^{l+1} for the next step (Eq. (4.8)).

$$h_{N(u)}^{(l+1)} = aggregate(\{h_u^l, \forall v \in N(u)\}) \quad (4.7)$$

$$h_u^{(l+1)} = \sigma(W \cdot concat(h_u^l, h_{N(u)}^{l+1})) \quad (4.8)$$

(2) GraphSAGE for aggregating environmental information
The original GraphSAGE model processes the information of each neighbor equally and cannot reflect the different influences of different neighbors on the embedding of the central node. Therefore, in the process of aggregating neighbor information in this paper, weights are set for each neighbor, and the attribute values e_{uv} of the edges are taken as the weights (Eq. (4.9)), so that the model

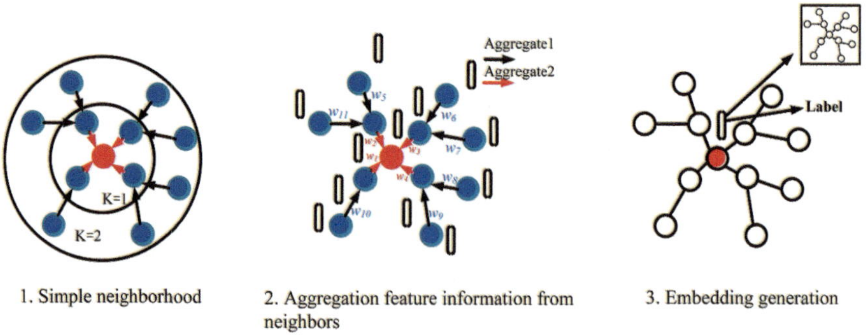

Fig. 4.7 The structure of GraphSAGE, reprinted from Zhang et al. (2024a), copyright 2025, with permission from Elsevier

can handle and capture the similar environmental information in space more effectively during the information transfer between graph nodes, and can better conform to the training of the directed weighted graph of geographic nodes with environmental consistency constraints. Since the neighbors of each node have been screened during the construction of the graph, when sampling the neighbors, the full neighbor set $N(u)$ of node u is selected instead of randomly sampling the neighbors of a fixed size. Furthermore, a self-loop was added for each node during the composition, so $h_{N(u)}^{l+1}$ contains the feature information of u itself. There is no need to perform the concat operation again, and the representation of node u in the next layer can be obtained for subsequent information transmission (Eq. (4.10)).

$$h_{N(u)}^{l+1} = aggregate(\{e_{uv} \cdot h_v^l, \forall v \in N(u)\}) \tag{4.9}$$

$$h_u^{(l+1)} = \sigma(W \cdot h_{N(u)}^{l+1}) \tag{4.10}$$

The GraphSAGE model designed in this study consists of three graph convolutional layers and two fully connected layers (Fig. 4.8). The GraphSAGE network is trained using supervised learning. The steps are as follows: First, use the Deep Graph Library (DGL) package in PyTorch to build the modified GraphSAGE model. Secondly, the dataset is randomly divided into 70% of the training samples and 30% of the test samples. The initial learning rate is set at 0.05, and the dynamic learning rate adjustment technique is adopted for optimization. To prevent the model from overfitting, we added a dropout layer to the node feature vectors and graph edges, with a ratio of 0.25. The number of training epoches is set at 2000. The Adaptive Moment Estimation (Adam) optimizer is adopted, and the binary cross-entropy is used as the loss function. The nonlinear fitting ability of the model was enhanced by using the Rectified Linear Unit (ReLU) activation function. Subsequently, after training and optimization, the GraphSAGE model generates representations of all nodes in the region. The fully connected classification layer is adopted to determine whether the nodes will form landslides, and the *softmax* function is used to evaluate the probability of landslides occurring at each node. Finally, the probabilities of each node are classified and mapped to the original geospatial domain to generate the final landslide susceptibility mapping.

4.4.3 Evaluation and Analysis of the Wide-Area Landslides Susceptibility Mapping

The model proposed in this paper was used to assess the susceptibility of landslides in the Bailong river basin, verifying the effectiveness and accuracy of the new method. Firstly, the accuracy analysis of the landslide susceptibility mapping results of the graph neural network with geographical environment consistency constraints

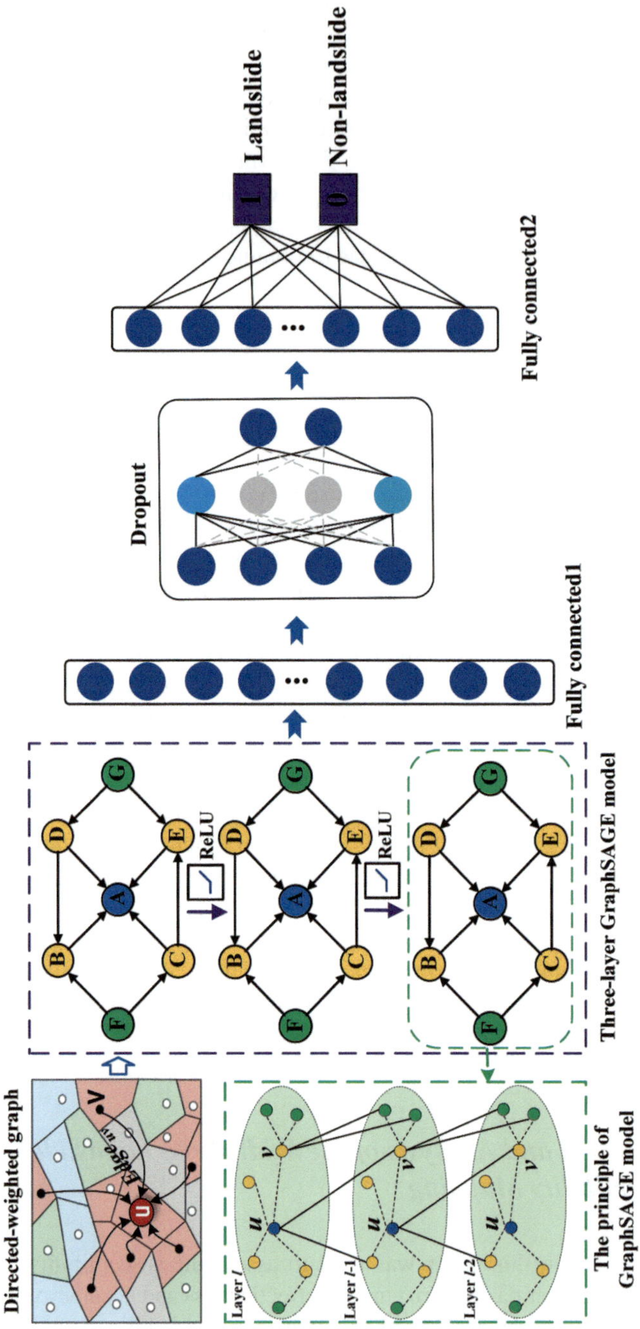

Fig. 4.8 The framework of the proposed model for LSM, reprinted from Zhang et al. (2024a), copyright 2025, with permission from Elsevier

proposed in this paper is conducted, including Accuracy, Precision, F1-score, Recall and ROC curve, etc. Then, different sampling strategies are designed. The mapping results of landslide susceptibility are compared and analyzed using the new method to verify the applicability of the new model and improve the reliability of graph neural networks for landslide susceptibility mapping. Finally, the landslide in the Bailong river basin is predicted using the method proposed in this paper and compared with the existing classical machine learning methods.

4.4.3.1 Model Performance Evaluation

(1) Comparison of model evaluation indicators

The confusion matrix is a common indicator for summarizing the classification and prediction results of the model in machine learning. Among them, True Positive (TP) represents the number of correctly predicted slippery samples, False Negative (FN) represents the number of correctly predicted non-slippery samples, and False Positive (FP) represents the number of landslide samples that were not correctly predicted, and True Negative (TN) represents the number of non-landslide samples that were not correctly predicted (Dou et al., 2020). The following indicators can be calculated using these values:

Accuracy is used to measure the proportion of samples that the model predicts correctly among the total samples. Its calculation formula is as follows:

$$Accuracy = \frac{TP + TN}{TP + FP + TN + FN} \tag{4.11}$$

Precision is used to measure the proportion of samples predicted by the model to be in the positive class that are actually in the positive class. Its calculation formula is as follows:

$$Precision = \frac{TP}{TP + FP} \tag{4.12}$$

Recall is used to measure the proportion of samples that the model correctly predicts as positive in the actual positive class. Its calculation formula is as follows:

$$Recall = \frac{TP}{TP + FN} \tag{4.13}$$

The F1 score (F-measure) is the weighted harmonic mean of precision and recall, which comprehensively considers the influence of precision and recall and can better reflect the comprehensive performance of the model. Its calculation formula is as follows:

$$F - measure = 2 \times \frac{\text{Precision} \times \text{Recall}}{\text{Precision} + \text{Recall}} \tag{4.14}$$

Furthermore, the ROC is a synthesis of the confusion matrices under different thresholds and is widely used in the mapping of landslide susceptibility. ROC takes the False Positive Rate (FPR) as the abscissa and the True Positive Rate (TPR) as the ordinate to reflect the continuous changes in data specificity and sensitivity. The AUC of ROC can visually reflect the results. The closer the value of AUC is to 1, the better the model effect is.

$$F - measure = 2 \times \frac{\text{Precision} \times \text{Recall}}{\text{Precision} + \text{Recall}} \tag{4.15}$$

$$TPR = \frac{TP}{TP + FN} \tag{4.16}$$

(2) Model performance evaluation results
Model accuracy mapping is crucial for measuring model performance. In order to evaluate the accuracy of the proposed model in the task of landslide susceptibility mapping, we conducted a precision analysis of the model. The dataset is divided into the training dataset and the test dataset in a ratio of 7:3. Figure 4.9 shows the results of the above evaluation indicators. The accuracy rate represents the ratio of the number of landslide samples correctly predicted by the model to the total number of landslide samples. It demonstrates the overall accuracy of the model for the occurrence of landslides. It can be known from Fig. 4.9 that the accuracy rates of the four models are all above 0.75, indicating that all four models can classify relatively accurately. However, for tasks like landslide susceptibility mapping, it is not a simple binary classification task. The classification situations of different categories should be analyzed in combination with the actual situation. For the landslide data with a true value of 1, only when the model truly learns the landslide characteristics can it make correct indications for the dangerous areas where landslides have not occurred. The recall rate reflects the model's ability to capture real landslide samples in the mapping of landslide susceptibility. It represents the ratio of the number of landslide samples correctly predicted by the model to the actual number of landslide samples. High recall rate means that the model can effectively identify real landslide situations. The recall rate of the model constructed in this paper is 0.91, indicating that the model can accurately identify landslides. Compared with the other three models, the model in this paper learns the characteristics of landslides more thoroughly. Accuracy is a key consideration in the mapping of landslide susceptibility, focusing on the prediction accuracy of the model for different categories. It represents the ratio of all the samples (including landslides and non-landslides) correctly predicted by the model to the total number of samples. Accuracy helps to understand the overall performance of the model in various categories. The accuracy of the model proposed in this paper is more than 2% higher than that of the other three models, indicating that the model has

a more comprehensive feature learning for landslides and non-landslides and can accurately classify landslide and non-landslide samples. The F1 index is a comprehensive reflection of accuracy and recall rate, and is used to measure the predictive ability of the model in different categories. This paper proposes that the F1 index of the model is above 0.8, indicating that the model can achieve a good balance between accuracy and recall rate when predicting landslides. This indicates that the model demonstrates good performance when predicting both positive and negative samples, does not favor either side, and has a high degree of credibility for predictions of different categories. Furthermore, the AUC index measures the classification ability and discrimination of the model for positive and negative samples under different thresholds. This paper proposes that the AUC value of the model is the highest, indicating that the model has a good discrimination between landslides and non-landslides, and can well explore the mapping relationship between landslides and influencing factors. It proves that the model constructed in this paper has the optimal prediction of landslide susceptibility, which provides important reference value for the auxiliary decision-making of landslide risks in the future.

Figure 4.10 shows the error distribution of the four models on the training and test datasets. It can be clearly seen from the figure that whether applied to the training dataset or the test dataset, the Error value of the model proposed in this paper is the smallest, and the Mean Squared Error (MSE) and Root Mean Squard Error (RMSE) are respectively (0.11, 0.34) and (0.14, 0.37). The results show that this model has strong robustness, good performance and high reliability.

(3) Comparative analysis of sampling strategies

Negative samples of landslides play an important role in the machine learning-based assessment of landslide susceptibility and can suppress the overestimation of landslide susceptibility by the model. This paper proposes a non-landslide sampling strategy considering the similarity of geographical environments. This strategy significantly improves the problem of low reliability of landslide susceptibility results caused by traditional sampling methods. The sample dataset is the basis for the construction of machine learning models. The training and test datasets, including both landslide and non-landslide ones, are crucial throughout the modeling process (Huang et al., 2021). Landslide samples are usually selected from historical landslide catalogues or remote sensing images, while non-landslide samples are usually not directly available (Zhu et al., 2022). At present, in most studies, non-landslide samples are obtained through random sampling (Chang et al., 2023). The credibility of negative samples collected using the random sampling method is unknown, and it is highly likely that potential landslides will be misclassified as negative samples, thereby hindering the model from learning the most effective features. Therefore, based on the geographical knowledge that "the more similar the geographical environment, the more similar the geographical features", it is believed that the location with a geographical environment similar to the original is very likely to be the location where future landslides will occur. Locations with geographical environments

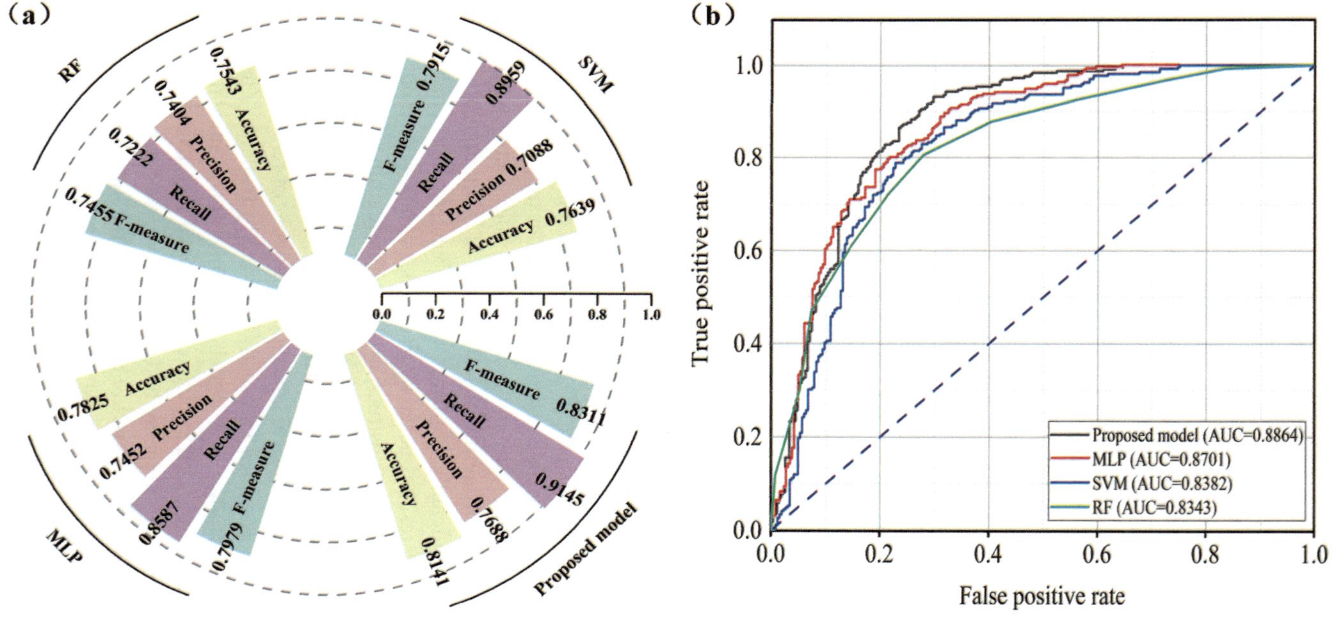

Fig. 4.9 Results of model evaluation metrics: **a** Accuracy, Precision, Recall and F-measure metrics, **b** AUC metrics, reprinted from Zhang et al. (2024a), copyright 2025, with permission from Elsevier

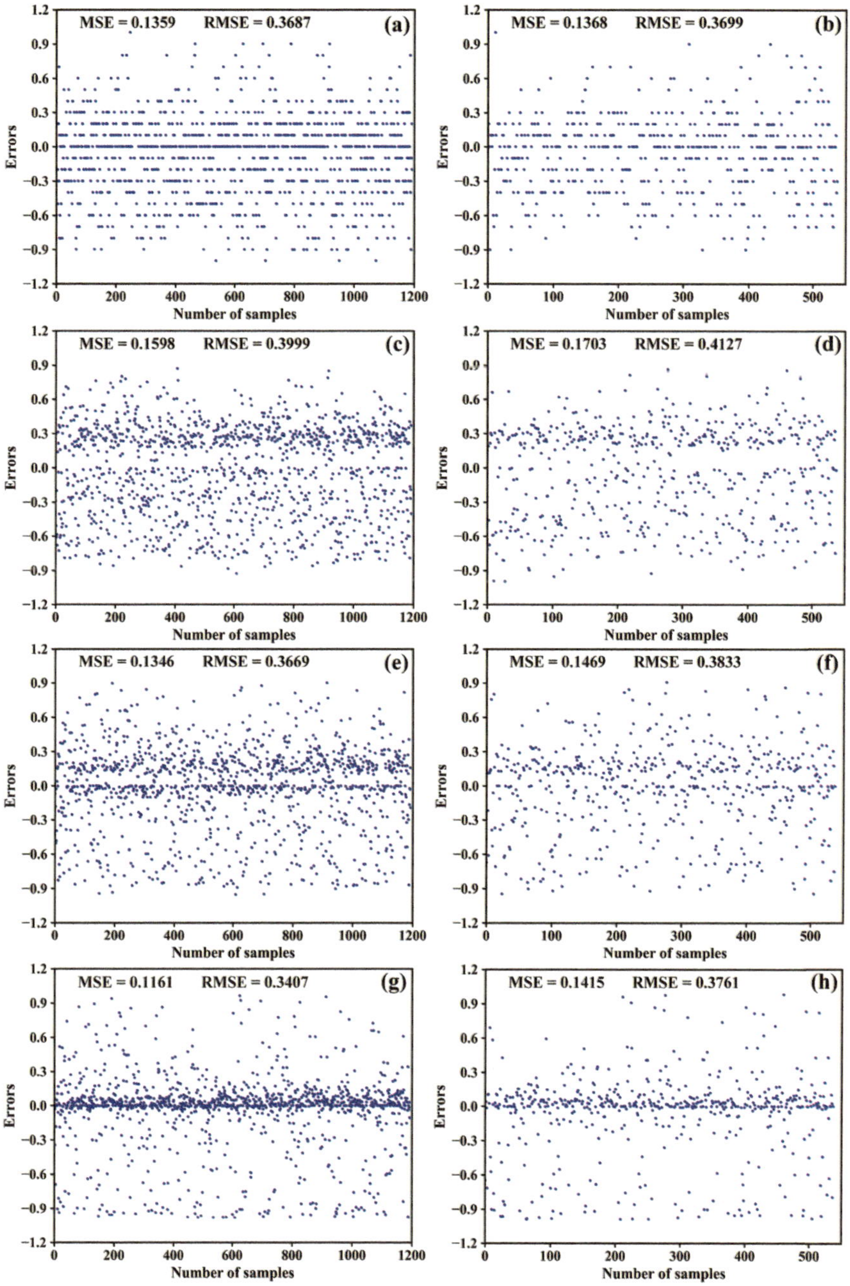

Fig. 4.10 The errors in MSE and RMSE for both training (left) and testing data (right): **a**, **b** RF, **c**, **d** SVM, **e**, **f** MLP, and **g**, **h** the proposed model, reprinted from Zhang et al. (2024a), copyright 2025, with permission from Elsevier

different from those of the positive samples are more likely to be negative samples. Based on this concept, non-landslide samples are selected to obtain a higher-quality dataset.

To verify the advantages of the sampling strategy based on similar geographical environments in this paper, a comparison is made with the conventional method of selecting samples by random sampling. The left side shows the mapping results of landslide susceptibility obtained by using the random sampling method, and the right side shows the mapping results of landslide susceptibility obtained by using the sampling method proposed in this paper (Fig. 4.11). It can be known from Fig. 4.12 that, compared with traditional sampling, the mapping results of landslide susceptibility obtained based on similar environment sampling are better, and the predicted area of high susceptibility areas is more consistent with the historical landslide distribution. In the absence of environmental constraints, all four models inevitably have missed judgments and overestimates of landslides. When the proposed sampling strategy was not adopted, the susceptibility areas in the RF model study area were widely distributed, and typical problems could not be highlighted. The distribution of very high and high susceptibility areas in the study area of the SVM model is concentrated, making it impossible to accurately and effectively identify the location of landslides. The MLP model has missed the judgment of historical landslides. In the circled areas of the graph, the location of the historical landslide points is predicted to be of low susceptibility. The model proposed in this paper has a wide and concentrated distribution of highly susceptibility areas and is unable to provide reference landslide location information. Under environmental constraints, the mapping results of landslide susceptibility of the four models are all good, proving that the proposed sampling strategy can enable the models to grasp the environmental characteristics of landslide disasters more comprehensively, reduce the influence of environmental differences on the models, and suppress the overestimation of landslide susceptibility. Furthermore, we also calculated the evaluation index results of the model under the two sampling methods. As shown in Fig. 4.12, the indicators of the model using environmental division are all higher than those of the model with random sampling. This indicates that determining the selected samples of various types of environments through environmental division and geographical environment similarity constraints can enable the model to learn the landslide formation characteristics of different environments. Enabling the model to deeply explore the potential correlations of landslide characteristics is conducive to summarizing and transmitting landslide characteristics in different environments and improving the reliability of prediction.

4.4.3.2 Analysis of Landslide Susceptibility Mapping

Under the data conditions of environmental constraints, we used four trained models, namely RF, SVM, MLP and GraphSAGE, to predict the landslide susceptibility in the Bailong River Basin. We assigned the predicted values to each slope unit to generate the landslide susceptibility mapping. The landslide susceptibility mapping

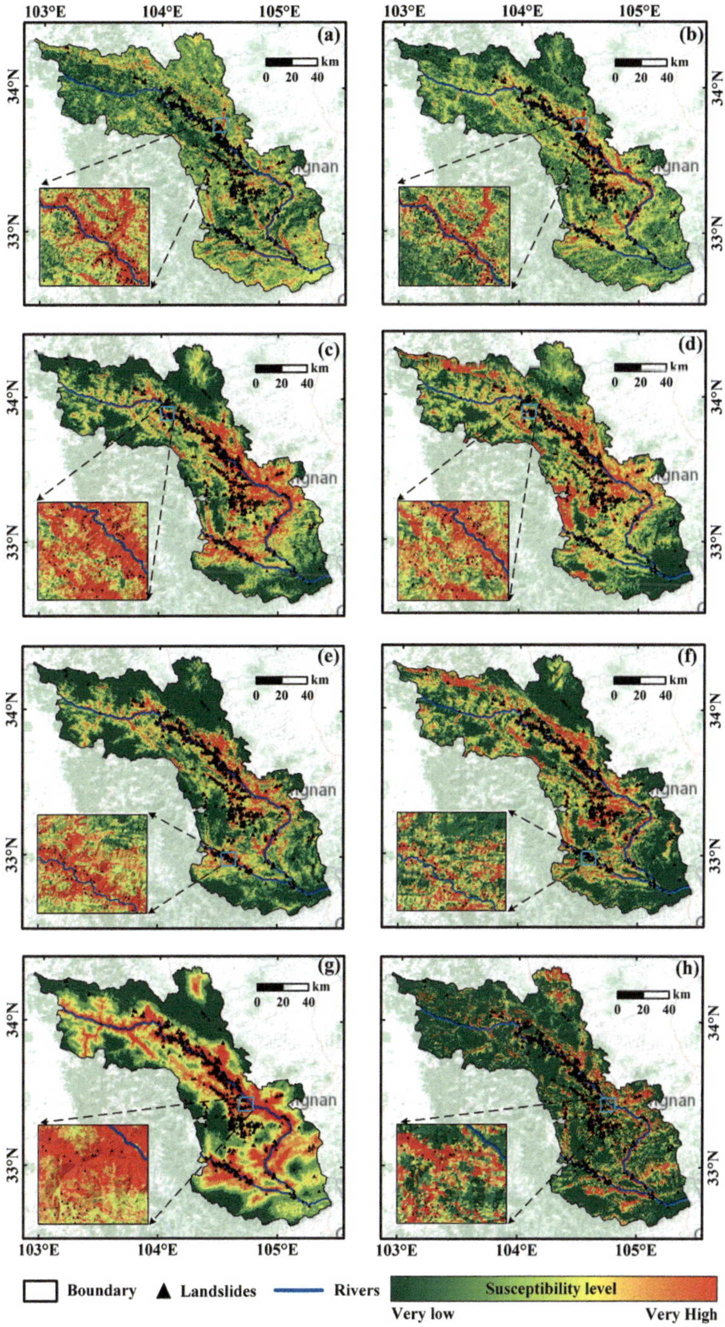

Fig. 4.11 Results of the LSA based on random sampling (left) and the proposed sampling method (right) using RF, SVM, MLP, and the proposed model, reprinted from Zhang et al. (2024a), copyright 2025, with permission from Elsevier

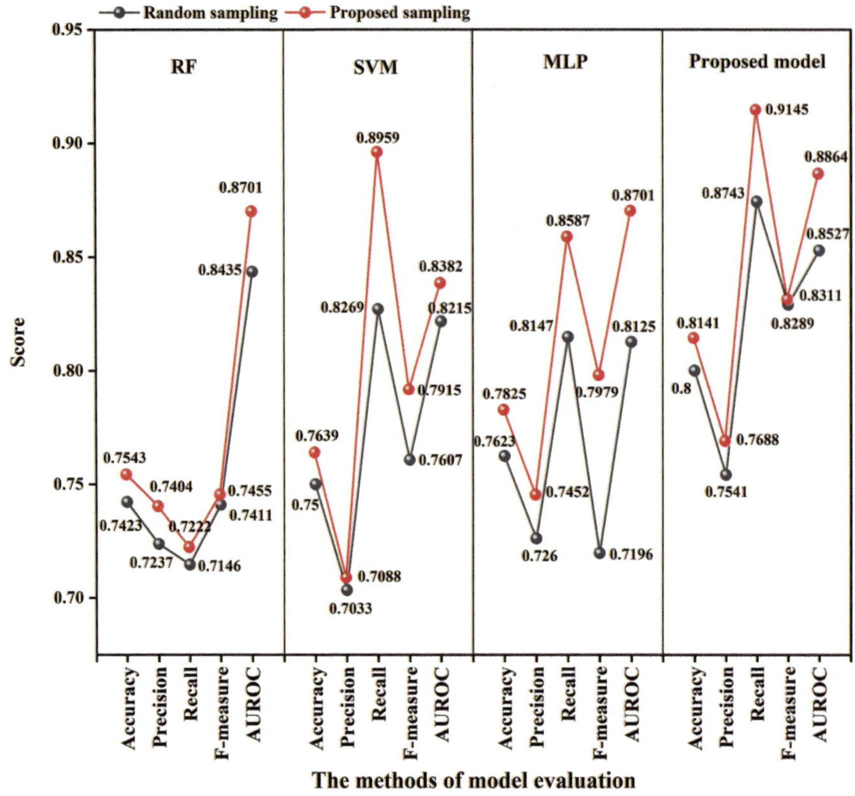

Fig. 4.12 The model evaluation index results of the model under different sampling strategies

is a visualization means that visually reflects the prediction results of the model (Wei et al., 2022). Figure 4.13 shows the mapping results of landslide susceptibility of four models. To quantify the prediction results of the models, the proportions of the four models in different susceptibility were statistically analyzed.

It can be known from Fig. 4.13 that the moderate-susceptibility region predicted by the RF model has the highest proportion among the four models and is widely distributed, accounting for approximately 23.19% of the study area. The classification results of susceptibility predicted by the SVM model were slightly better than those of the RF model. However, the distribution range of very high and high susceptibility areas was wide, accounting for 11.33% and 18.27% of the study area respectively. In the MLP prediction results, the regions divided by various categories have improved compared with RF and SVM. In the prediction results of the method proposed in this paper, the proportions of very high and very low susceptibility areas are the highest, accounting for 13.21% and 57.61% of the study area respectively, and these areas are consistent with the locations of most historical landslides.

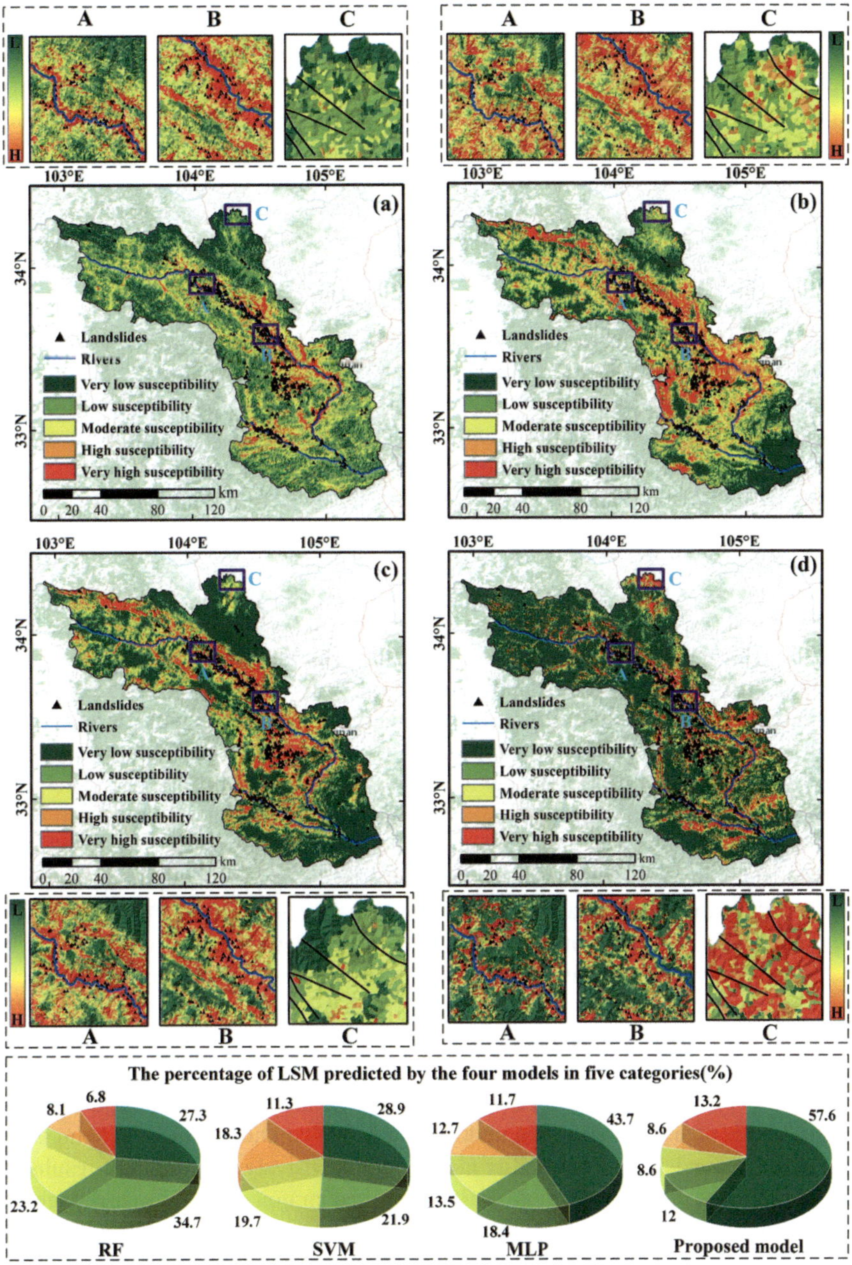

Fig. 4.13 Landslide susceptibility mapping: **a** RF, **b** SVM, **c** MLP, **d** Proposed model, reprinted from Zhang et al. (2024a), copyright 2025, with permission from Elsevier

The RF model learns the landslide characteristics by constructing multiple decision trees (Catani et al., 2013); The SVM model maps the landslide features to the high-dimensional space by introducing a kernel function, making the features linearly separable in the high-dimensional space (Marjanović et al., 2011); The MLP model learns landslide characteristics through the stacking of multiple layers of neurons (Ermini et al., 2005). Although these three machine learning methods have their own advantages in learning the characteristics of landslides, none of them take into account the correlation between landslides and the disaster-prone environment they are in. They only use the features of independent samples to learn the feature distribution of landslides and perform well in local performance. However, when facing complex geographical environments, they produce unbalanced prediction results. In contrast, the GraphSAGE model proposed in this paper adopts a graph structure to display and express the correlation between landslides and disaster-prone environments. It transmits node feature information through nodes and edges to achieve cross-spatial feature transmission. It can associate landslide points with similar disaster-prone environments on a larger scale and describe the spatial relationship between landslides more effectively. So as to predict the susceptibility of landslides more accurately.

In the mapping of landslide susceptibility mapping, the performance of the model is often restricted due to the limited sample dataset. In this paper, under the same data conditions, compared with the other three models, the model proposed in this paper has the best mapping result of landslide susceptibility mapping. The reason lies in that the GraphSAGE model adopted in this paper can transfer the landslide characteristics learned in the area with landslide samples to the environment without landslide samples but similar, greatly reducing the missed alarm rate of the model and helping to improve the susceptibility in the area without landslide samples. As shown in Area A and Area B of Fig. 4.13, the model proposed in this paper is high susceptibility in prediction, closer to the distribution of historical landslides, and provides a more detailed division of landslide areas. Furthermore, as shown in area C of Fig. 4.13 this area has no historical landslide point data and is relatively close to the fault. The proposed model predicts it as a high susceptibility area, while the other three models predict it as a low susceptibility area. This indicates that the model can achieve reliable mapping of landslide susceptibility and is capable of exploring the relationship between landslides and landslide-prone environments across space and effectively. This further proves that the model we proposed can learn the most accurate and effective landslide characteristics even under limited data conditions, thereby being able to predict the landslide susceptibility more comprehensively. Although the results of the four models are different, there are still certain commonalities. The susceptibility evaluation results of the four models show that the very high susceptibility area and the high susceptibility area are mainly located in the middle of the Bailong river basin, while the northwest is a large area of the very low susceptibility area and the low susceptibility area, and only a small area of the moderate susceptibility area, the high susceptibility area and the very high susceptibility area are distributed in mountainous landforms. The geological environment in the eastern and southern parts of the Bailong river basin is relatively complex, the water system is well-developed, the seismic movement is active and the rainfall is relatively large. The very high and high susceptibility areas are mainly distributed along both sides of the river.

To further verify the reliability of the prediction results of the model in this paper, field investigations can be conducted to verify whether the high susceptibility areas predicted by the model constructed in this paper have landslide disasters. To better present the prediction results, a landslide area that does not appear in the landslide catalog can be selected to evaluate the accuracy of the model. Through field investigation, two typical areas were selected in this paper. Figure 4.14 shows the landslide susceptibility mapping results, Google Images and unmanned aerial vehicle field investigation maps of the two areas. It was learned from field investigations that both landslide incidents occurred in June 2023. For region A, the model proposed in this paper predicts that this region is a very high susceptibility region, while the SVM model predicts that this region is a high susceptibility region, and the MLP and RF

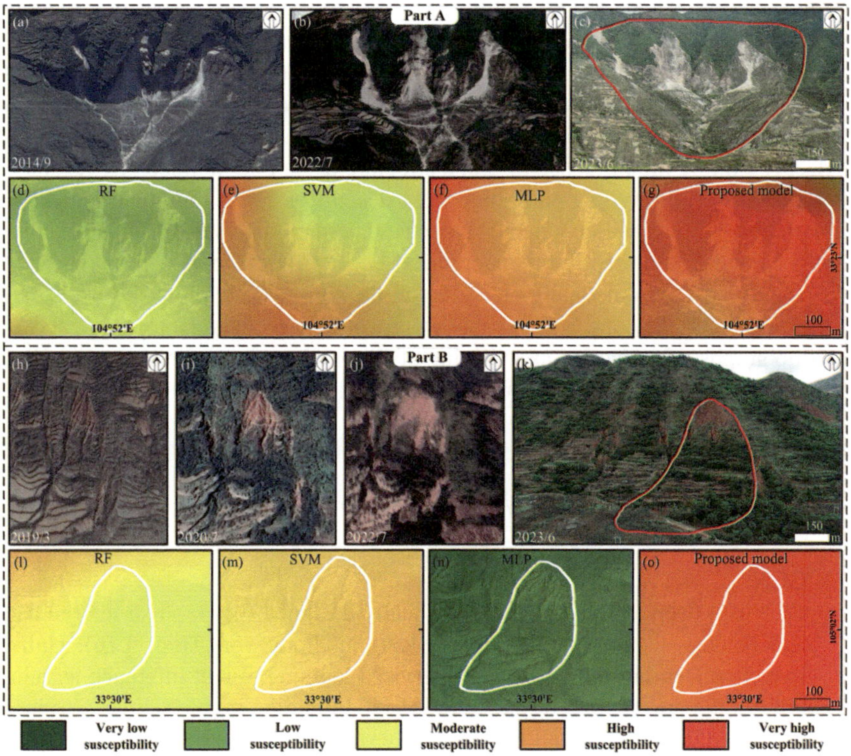

Fig. 4.14 Analysis of model prediction results based on Google satellite image and fieldwork: **a**, **b** Google satellite image, **c** fieldwork photos, **d** RF, **e** SVM, **f** MLP, and **g** the proposed model for part A; **h**, **i** Google satellite image, **k** fieldwork photos, **l** RF, **m** SVM, **n** MLP, and **o** the proposed model for part B. The red and white lines represent the landslide boundaries, reprinted from Zhang et al. (2024a), copyright 2025, with permission from Elsevier

models predict that this region is a low susceptibility region and a moderate susceptibility region respectively. For region B, the model proposed in this paper predicts that this region is a very high susceptibility region, while the SVM and MLP models predict that this region is a high susceptibility region, and the RF model predicts that this region is a moderate susceptibility region. The model proposed in this paper can efficiently predict the location of landslides and is of great significance for landslide prevention and management.

4.5 Construction of Landslide Susceptibility Mapping Model Based on Graph-Transformer

This chapter mainly studies how to optimize the application of graph neural networks in the mapping of landslide susceptibility by introducing the Transformer module, thereby achieving landslide susceptibility prediction on a larger scale. Considering that the acquisition of landslide data and the construction and processing of environmental similarity graph structures have been elaborated in detail in Sects. 4.1 and 4.4, the relevant contents will not be repeated in this chapter. Furthermore, in order to deeply explore the influence of graph structure design on the mapping results of landslide susceptibility mapping, this paper designs a spatial topological relationship graph structure and conducts a systematic comparative study on the mapping results of landslide susceptibility mapping based on different graph structures. Next, this chapter will systematically elaborate on the following key links: introduction to traditional Graph neural network models, construction of spatial topological relationship graphs, construction and training of Graph-Transformer models, as well as evaluation and analysis of experimental results (Fig. 4.15).

4.5.1 Traditional Graph Neural Network Model

(1) GCN

GCN is a deep learning model specifically designed for graph-structured data and belongs to the direct inference learning framework (Fig. 4.16). Unlike inductive learning in traditional supervised learning, direct inference learning utilizes the information of the entire graph during the training process to infer unlabeled data points, and is particularly suitable for cases with sparse labels. The main idea of GCN is to integrate the information of neighboring nodes by means of the adjacency matrix of the graph and the convolution operation, and then complete the representation learning of the graph structure (Kipf and Welling, 2017). Specifically, each node achieves the update of features by collecting and summarizing the information of its adjacent nodes. This process is accomplished through weight weighting, where the weights usually take into

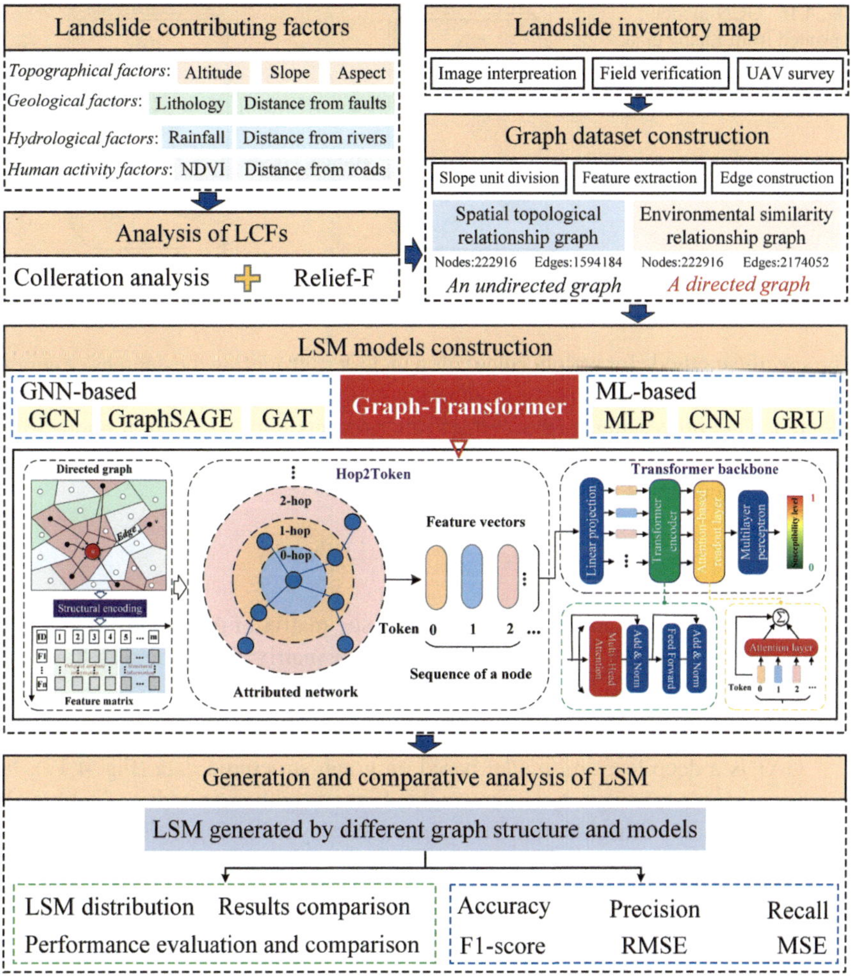

Fig. 4.15 Overall workflow of our approach, reprinted from Zhang et al. (2024b), copyright 2025, with permission from IEEE

account the degrees of neighboring nodes to ensure a reasonable allocation of the contributions of each neighboring node. In GCN, information fusion can be accomplished through the following steps:

Construct the adjacency matrix: First, generate the adjacency matrix of the graph, which is used to describe the connection relationship between nodes.

Normalization processing: Perform a normalization operation on the adjacency matrix to ensure that the influence of each node on its neighboring nodes is evenly distributed.

Convolution operation: By multiplying with the weight matrix, the information of neighboring nodes is integrated into the feature representation of the current node.

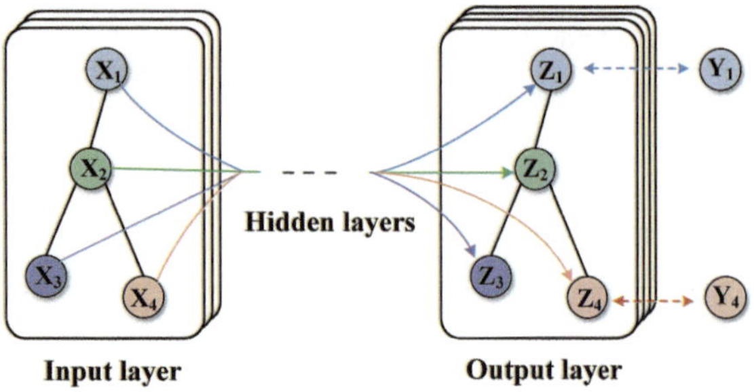

Fig. 4.16 GCN model, reprinted from Zhang et al. (2024b), copyright 2025, with permission from IEEE

The specific methods for weight calculation include considering the degrees of neighboring nodes, and symmetric normalization or Laplacian normalization is usually adopted. This processing method effectively prevents high-connection nodes from causing excessive interference to the results, thereby improving the stability and prediction accuracy of the model.

$$H^{(l+1)} = \sigma(\widetilde{D}^{-\frac{1}{2}}\widetilde{A}\widetilde{D}^{-\frac{1}{2}}H^{(l)}W^{(l)}) \tag{4.17}$$

In the formula, $H^{(l)}$represents the characteristic matrix of the *l-th* layer, and $H^{(0)} = X$; $\widetilde{D} = D + \text{I}$; $\widetilde{A} = A + I$, where I is the identity matrix; $W^{(l}$ is the weight matrix of the *l-th* layer filter; Represent the nonlinear activation function.

(2) GAT
GAT is a deep learning model based on graph-structured data (Fig. 4.17). Its main idea is to assign dynamic weights to the neighboring nodes of each node with the help of the attention mechanism, thereby achieving differentiated integration of the information of neighboring nodes (Veličković et al., 2018). In this way, the model can focus more on the neighbor nodes that are more important to the current node, thereby generating a more distinguishable representation of node features. These characteristics enable GAT to perform well in various graph-related tasks (such as node classification, graph classification and link prediction), and become one of the classic methods in the field of graph neural networks. In GAT, the specific implementation process of the attention mechanism can be decomposed into the following several steps:

Calculate the attention coefficient: For each node and its corresponding neighboring nodes, it is first necessary to determine the attention coefficient between them to quantify the importance of the neighboring nodes to the current node. Specifically, for the central node i and its neighboring node j, GAT calculates the attention weight α_{ij} between them by the following method:

$$\alpha_{ij} = softmax\left(\frac{\vec{a}\,(Wh_i)^{\top}Wh_j}{\sqrt{d}}\right) \tag{4.18}$$

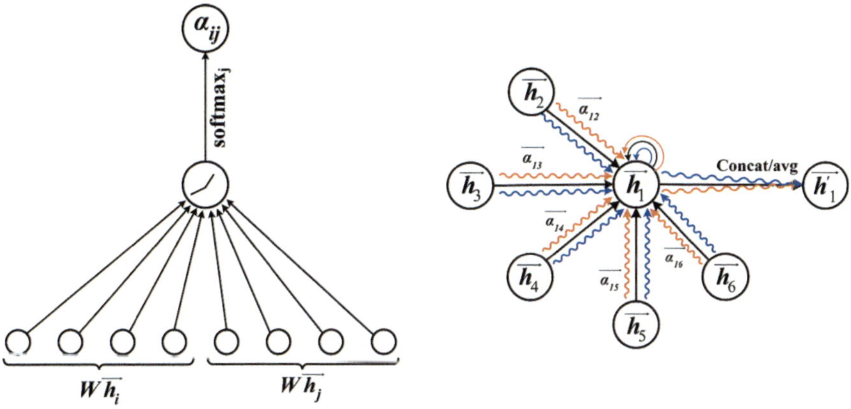

Fig. 4.17 GAT model, reprinted from Zhang et al. (2024b), copyright 2025, with permission from IEEE

Among them, W is a trainable linear transformation matrix, which is used to project the node features into a higher-dimensional space; $\vec{a}$ is another learnable parameter vector used to calculate the attention score; h_i and h_j represent the initial characteristics of node i and node j respectively; $\sqrt{d}$ is a scaling factor used to avoid numerical instability.

In practical operation, linear transformation is usually performed on Wh_i and Wh_j first, and then nonlinearity is introduced through the LeakyReLU activation function:

$$e_{ij} = LeakyReLU\left(\vec{a}\,(Whi)^{\top} Whj\right) \tag{4.19}$$

Finally, the attention scores of all neighboring nodes are normalized through the Softmax function to obtain the final attention coefficient α_{ij}.

Weighted aggregation of neighbor features: After obtaining the attention coefficients, the model will perform a weighted summary of the feature representations of neighbor nodes based on these weights:

$$h_i' = \sigma\Big(\sum_{j \in N(i)} \alpha_{ij} Wh_j\Big) \tag{4.20}$$

Among them, $N(i)$ represents the set of neighbor nodes of node i; W is a learnable weight matrix; σ is a nonlinear activation function. In this way, the new representation h_i' of each node will pay more attention to those neighboring nodes that are more important to the current task.

Multi-head attention mechanism: To enhance the model's expressive power, GAT has introduced a multi-head attention mechanism. Specifically, the model can be calculated simultaneously through multiple independent attention heads, and each attention head uses a different parameter matrix W_k and $\vec{a}_k$ sum to determine the attention weight. Ultimately, the output results of all the attention heads will be stitched together and integrated through linear transformation:

$$h_i' = concat(\{h_i^{(1)}, h_i^{(2)}, \ldots, h_i^{(k)}\})W\prime \tag{4.21}$$

Among them, K represents the number of attention heads; W 'is another learnable linear transformation matrix.

Superimpose pooling layers or fully connected layers: After generating new node representations, GAT can further add global pooling layers (such as average pooling or maximum pooling) or fully connected layers based on the requirements of specific tasks, thereby obtaining graph-level feature representations.

$$g = Pool(\{h_1', h_2', \ldots, h_N'\}) \tag{4.22}$$

Among them, Pool can adopt Mean Pooling, Max Pooling or other custom aggregation methods.

4.5.2 Construction of Spatial Topological Relationship Diagram

In this study, in order to construct the graph structure based on spatial topological relationships, the spatial analysis function of ArcGIS software was adopted and combined with the polygon neighborhood function in its "Analysis Tool" for operation (Zhang et al., 2024b). Specifically, first of all, the research area of the Bailong river basin is divided into several slope units as the basic analysis units. Each slope unit represents a specific spatial area and has certain geographical characteristics and slide-related attributes. Subsequently, through the polygonal neighborhood analysis algorithm of ArcGIS, the spatial adjacency relationship of each slope unit is calculated, thereby extracting the information of all neighborhood elements directly adjacent to the slope unit. During this process, special attention is paid to ensuring the accuracy and integrity of spatial adjacency relations. On the one hand, by preprocessing the data of the research area, the spatial consistency and quality of the input data are ensured; On the other hand, in the analysis of polygonal neighborhoods, calculations are carried out strictly in accordance with spatial topological rules to avoid the omission or error of adjacency relations caused by blurred boundaries or data errors. Based on the above steps, a graph structure based on spatial topological relations was finally generated (Fig. 4.18).

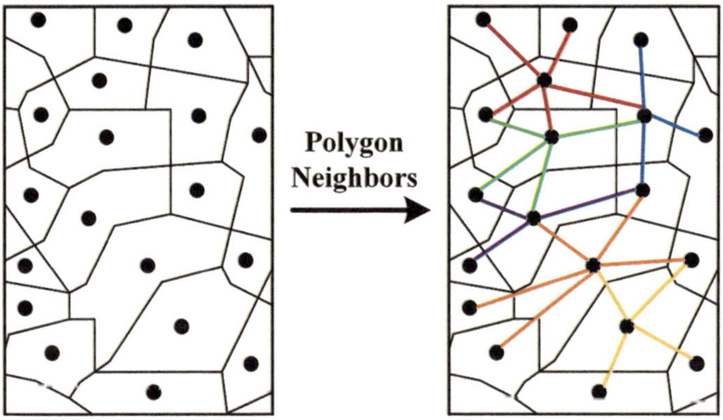

Fig. 4.18 Spatial topological relationship graph structure, reprinted from Zhang et al. (2024b), copyright 2025, with permission from IEEE

4.5.3 Construction of Graph-Transformer Landslide Susceptibility Mapping Model

4.5.3.1 Transformer

In 2017, the Google research team proposed the Transformer model, and this technology quickly became a milestone innovation in the field of natural language processing (Vaswani et al., 2017). It has been widely applied in numerous tasks, such as language translation, text generation, question-answering systems, and text summarization, etc. The Transformer model is composed of two major parts: the encoder and the decoder. Among them, the role of the encoder is to map the input sequence to the high-dimensional semantic representation space, while the decoder generates the target output sequence based on these semantic representations. Both the encoder and the decoder are composed of multiple identical modules superimposed, and each module contains two main sub-layers: the Multi-Head self-attention mechanism (MHA) and the Feedforward Neural Network (FFN).

The self-attention mechanism is a technique used to measure the degree of correlation between different positions in the input sequence. This method generates a global semantic expression by constructing the relationship between each element in the sequence and all the other elements. Compared with traditional loop structures (such as RNN), the self-attention mechanism can directly capture long-distance dependencies without the need to explicitly loop or slide the window operations on the sequence. The core idea of the self-attention mechanism is to compare the information of each position with that of all other positions and generate weight coefficients based on similarity. These weights determine the degree of attention that the current position pays to other position information, thereby achieving efficient global dependency modeling.

$$Attention(Q, K, V) = sofmax\left(\frac{QK^T}{\sqrt{d_k}}\right)V \quad (4.23)$$

In the formula, Q, K, and V represent the query vector, key vector, and value vector respectively.

To enhance the expressive ability of the model, the Transformer introduces a multi-head self-attention mechanism. Specifically, the input sequence will be projected simultaneously onto multiple independent query, key and value spaces, each corresponding to a unique attention head. These attention heads can be computed in parallel and eventually the results can be concatenated to generate a richer semantic representation.

$$head_i = Attention(QW_i^Q, KW_i^K, VW_i^V) \quad (4.24)$$

$$MultiHead(Q, K, V) = Concat(head_1, head_2, ..., head_h)W^O \quad (4.25)$$

In the formula, *MultiHead* represents the multi-head attention mechanism, *Concat* indicates the concatenation of the results of the multi-head attention mechanism, W^O is the weight matrix of the linear transformation of the concatenation results, while W_i^Q, W_i^K and W_i^V are respectively the weight matrices of the query, key, and value corresponding to each head in the multi-head attention mechanism.

The feedforward neural network consists of two linear transformation layers, and nonlinear activation functions (such as ReLU or GELU) are embedded between these two layers. Their role is to generate independent feature representations for each position, thereby further enriching the semantic expression ability of the model. Through this design, the Transformer model is not only capable of capturing global dependencies, but also has the ability to extract local context information.

$$Sublayer(x) = FeedForward(x) \; or \; MultiHead(x, x, x) \quad (4.26)$$

$$LayerNorm(x + Sublayer(x)) = LayerNorm(x + dropout(Sublayer(x))) \quad (4.27)$$

In the formula, *LayerNorm* represents the layer normalization operation, and *Sublayer* represents the sub-layer. Each sub-layer can be *FeedForward* or *MultiHead(x,x,x)*, and a residual connection is added.

4.5.3.2 Graph-Transformer

To solve the problem of the limited feature extraction ability of the landslide susceptibility mapping model based on Graph neural networks, this paper proposes a hybrid model Graph-Transformer for the landslide susceptibility mapping task, which improves the spatial prediction accuracy of landslides by extracting global information through the Transformer. The main innovation point of this study lies in achieving an efficient fusion of the characteristics of the landslide-prone environment and global information by combining the environmental similarity map with the Transformer model. Specifically, Graph is used to construct the spatial correlation structure of the landslide-prone environment and capture the interaction relationship of geographical elements within the local area. The Transformer module extracts the global feature information of landslide data through the self-attention mechanism, thereby comprehensively analyzing the influence of complex terrain conditions on the occurrence of landslides on a larger scale. The specific architecture of Graph-Transformer is shown in Fig. 4.19, which contains four core modules: The structure coding module considering the characteristics of graph structure, the Hop2Token module converting node domain information to sequence information, the Transformer encoder module, and the attention-based readout funcation module considering the influence of different neighboring nodes.

Structural encoding: In the graph structure, in addition to the attribute information of the nodes themselves, the structural information between nodes is also equally important for the learning of the model. Adjacent nodes usually have similar positional characteristics, while nodes that are farther apart exhibit different positional characteristics. The graph Laplacian matrix (Δ) can comprehensively reflect the positional relationship and connection strength between nodes in the graph, and the feature vectors obtained through its decomposition can accurately describe the positional information and feature distribution of nodes. Based on this, in this paper, the feature vectors obtained from the decomposition of the graph Laplacian matrix are used to represent the structural position information of the graph, and they are concatenated with the original attribute information matrix (X') of the graph structure to construct the feature matrix (X).

$$\Delta = I - D^{-1/2} A D^{-1/2} = U^T \Lambda U \tag{4.28}$$

In the formula, A represents the n × n adjacency matrix, D is the degree matrix, Λ and U corresponds to its eigenvalue and eigenvector respectively.

$$X = X' \| U \tag{4.29}$$

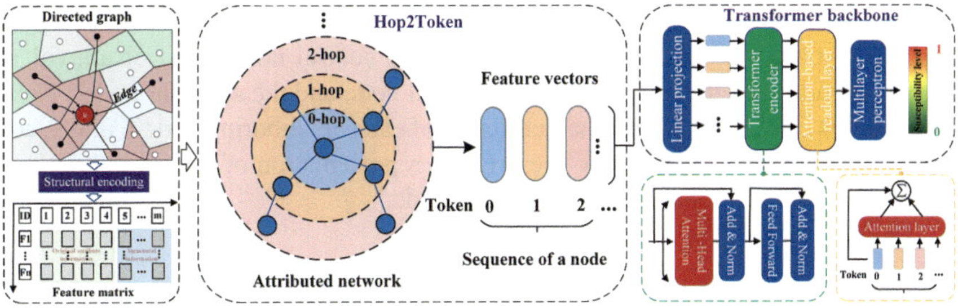

Fig. 4.19 Framework of graph–transformer, reprinted from Zhang et al. (2024b), copyright 2025, with permission from IEEE

In the formula, X is the feature matrix and X' is the original attribute information matrix.

Hop2Token: After obtaining the feature matrix of the graph structure, the feature information of neighbors of different orders of the central node can be aggregated by multiplying it with the adjacency matrix. Assuming that the maximum order of the aggregated neighbors is K, the aggregated neighbor information is A^KX. The specific implementation process is shown as Algorithm 1.

Algorithm 1 Hop2Token
Input: Normalized adjacency matrix A; Feature matrix X; Propagation step K
Output: Sequences of all nodes X_G
1: **for** $k = 0$ to K **do**
2: **for** $i = 0$ to n **do**
3: $X_G[i, k] = X[i]$;
4: **end for**
5: **X** = **AX**;
6: **end for**
7: **return** Sequences of all nodes X_G;;

X_G represents the feature information matrix of each node after aggregating the information of K-th neighbor nodes. For one of the nodes u, the aggregated $K - th$ neighbor information can be expressed as $S_u = (x_u^0, x_u^1, ..., x_u^k)$. Then use a learnable linear projection S_u to map the hidden dimensions to the Transformer d_m layer:

$$Z_u^{(0)} = \left[x_u^0 E; x_u^1 E; ...; x_u^K E\right] \tag{4.30}$$

In the formula, E $\in R^{d' \times d_m}$, $Z_u^{(0)} \in R^{(K+1) \times d_m}$ $Z_u^{(0)} \in R^{(K+1) \times d_m}$.

Transformer encoder: Send the projected sequence into the Transformer encoder. LayerNorm is applied before each module, while FFN contains two linear layers and uses GELU as the nonlinear activation function:

$$Z_u'^{(l)} = MSA\left(LN\left(Z_u^{(l-1)}\right)\right) + Z_u^{(l-1)} \tag{4.31}$$

$$Z_u^{(l)} = FFN\left(LN\left(Z_u'^{(l)}\right)\right) + Z_u^{(l)} \tag{4.32}$$

Attention-based readout funcation: Considering the different influences of neighboring nodes on the central node, this paper proposes an attention-based readout function to obtain the final representation of the central node. First of all, calculate the attention coefficient between the central node and the adjacent nodes. Next, multiply this coefficient by the characteristics of each neighboring node. Finally, add the result of the multiplication to the original representation of the central node to obtain the final representation of the central node.

$$\alpha_k = \frac{exp((Z_0 \| Z_k) W_a^T)}{\sum_{i=1}^{K} exp((Z_0 \| Z_i) W_a^T)} \tag{4.33}$$

In the formula, Z_0 is the node itself, Z_k is its k-hop representation, $W_a \in R^{1\times 2d_m}$ represents the learnable projection, i = 1,… K. The final aggregated representation of the nodes is as follows:

$$Z_{out} = Z_0 + \sum_{k=1}^{K} \alpha_k Z_k \tag{4.34}$$

4.5.4 Evaluation and Analysis of the Wide-Area Landslides Susceptibility Mapping

To verify the validity and accuracy of the Graph-Transformer model based on the attention mechanism proposed in this paper in the mapping of landslide susceptibility in the Bailong river basin, a systematic experimental scheme was designed in this study. Firstly, by comprehensively evaluating the predictive performance of the model, the accuracy analysis of the landslide susceptibility mapping method proposed in this paper is conducted to ensure its feasibility and effectiveness. Secondly, in the experimental scenario of the Bailong river basin, the method proposed in this paper was adopted for landslide prediction, and a comprehensive comparative analysis was conducted with the existing classical graph neural network methods to further prove the significant advantages of the method proposed in this paper in terms of model performance and prediction accuracy. Finally, an in-depth comparison was made of the mapping results of landslide susceptibility under different graph structures to systematically verify the scientificity and applicability of the graph structures designed in this paper.

4.5.4.1 Model Performance Evaluation

To reflect the influence of different graph structures on the model performance and quantify the differences in the learning abilities of the models, this paper calculates the evaluation indicators of four models under different graph structures. It can be seen from Table 4.3 that under the structure of the environmental constraint graph, the evaluation indicators of the four models (including Accuracy, Precision, Recall, F1 and AUC) are all higher than their performances under the structure of the spatial relationship graph. Among them, the GCN model increased by 20.65%, 15.61%, 14.34%, 18.77% and 19.22% respectively, and the GraphSAGE model increased by 13.49%, 12.68%, 12.07%, 10.55% and 13.89% respectively. The GAT models increased by 25.12%, 21.98%, 27.71%, 24.65% and 27.52% respectively, and the proposed models increased by 9.41%, 7.06%, 11.03%, 9.35% and 8.90% respectively. This indicates that the graph structure with environmental constraints promotes the learning of the model, which can help the model learn the most effective features and effectively improve the prediction accuracy of the model. Furthermore, under the same graph structure, the model proposed in this paper is superior to the other three models. Among them, the evaluation index results such as Accuracy, Precision, Recall, F1 and AUC calculated, under the spatial relationship graph structure, they have increased by at least 5.91%, 11.60%, 3.24%, 3.95% and 6.73% respectively; Under the structure of the environmental constraint graph, they have increased by at least 0.63%, 6.03%, 2.28%, 0.49% and 2.05% respectively. This indicates that the model proposed in this paper has a strong feature learning ability, can extract features more comprehensively, deeply reveal the potential relationship between landslides and influencing factors, and achieve an effective association between landslides and disaster-prone environments.

Statistical error distribution is an important indicator for evaluating the predictive performance of a model. It can visually reflect the accuracy and deviation degree of the model during the prediction process. To comprehensively evaluate the performance of the model, in this paper, under different graph structure conditions, the MSE and RMSE of four models on the test dataset were calculated. It can be seen from Fig. 4.20 that in the graph structure based on spatial relations, the error performance of the GCN model is relatively poor, and its MSE and RMSE reach 0.19 and 0.44 respectively. In contrast, the model proposed in this paper shows a lower error level, with the corresponding MSE and RMSE being 0.14 and 0.35 respectively. This indicates that, considering only the spatial relationship, the model in this paper can extract the useful information in the data more effectively, thereby achieving higher prediction accuracy. Further analysis reveals that under the graph structure with environmental constraints introduced, the error indicators of each model have decreased to a certain extent. Among them, the error of the GAT model is still relatively high, with its MSE and RMSE being 0.10 and 0.32 respectively. The model proposed in this paper achieved the best performance under this condition, with its MSE and RMSE dropping to 0.05 and 0.24 respectively. These results indicate that the environmental constraint graph structure can, to a certain extent, make up for the deficiency of relying solely on the spatial relationship graph structure, significantly improve the predictive ability of the model, and effectively reduce the possibility of misjudgment and missed judgment. Overall, the above experimental results fully demonstrate the positive impact of the environmental constraint graph structure on the model performance, and can effectively improve the prediction accuracy and generalization ability

Table 4.3 Evaluation index results under different graph structures and models, reprinted from Zhang et al. (2024b), with permission from IEEE

	Spatial topological relationship graph structure				Environmental similarity relationship graph structure			
	GCN	GraphSAGE	GAT	Proposed	GCN	GraphSAGE	GAT	Proposed
Accuracy	0.71	0.75	0.69	0.79	0.86	0.85	0.86	0.87
Precision	0.68	0.70	0.65	0.78	0.79	0.79	0.79	0.84
Recall	0.78	0.81	0.71	0.84	0.90	0.91	0.91	0.93
F1	0.73	0.77	0.68	0.80	0.87	0.85	0.85	0.87
AUC	0.77	0.81	0.71	0.86	0.922	0.925	0.90	0.94

of the model. Meanwhile, the model designed in this paper shows strong stability and excellent predictive ability under both graph structure conditions, providing strong support for its practical application in the field of landslide susceptibility mapping.

4.5.4.2 Analysis of Landslide Susceptibility Mapping

To visually reflect the rationality of the constructed graph structure and the performance of the model, the same dataset is applied to different graph structures and models, and the advantages and disadvantages of different graph structures and models are measured by comparing the prediction results. Figure 4.21 shows the prediction results of landslide susceptibility obtained under different graph structures and models. Furthermore, in order to quantify the differences in results under different graph structures and models, this paper conducts statistical analysis on the proportion of each sensitive area in the landslide susceptibility mapping and the distribution proportion of historical landslide points in each susceptibility area.

Under the structure of the spatial relationship diagram, the areas with moderate and high susceptibility predicted by the GCN model are widely distributed, accounting for 19.59% and 19.48% of the study area respectively. However, only 44.74% of the landslide sites are located in very high susceptibility areas, making it difficult to provide reliable landslide reference information. Approximately 66.36% of the landslides in the GraphSAGE model are located in very high susceptibility areas, but about 15% of the landslide points still exist in moderate and low susceptibility areas. The prediction results have a certain false alarm. In the prediction results of the GAT model, most of the susceptibility areas are of low susceptibility (29.92%), and the distribution of the very high susceptibility areas is discrete. Only 49.77% of the landslide points overlap with this area. In the prediction results of the proposed model, very low and very high susceptibility areas are distributed in large areas, accounting for 57.80% and 11.26% of the study area respectively. Approximately 62.01% of the landslide points are located within the very high susceptibility areas. In contrast, under the graph structure with environmental similarity constraints, the prediction results of landslide susceptibility mapping of the four models have been greatly improved, and the predicted high susceptibility areas are more consistent with the actual distribution. Among them, the historical landslide points accounted for 92.11%, 74.49%, 78.60% and 89.70% respectively in the very high susceptibility areas of the four models. Meanwhile, very high and high susceptibility areas show a trend of distribution along rivers. This indicates that the graph constructed based on the environmental similarity relationship is more reasonable, can fully correlate similar landslide-prone environments, can more effectively explore the correlation between landslides-prone environments, and the landslide susceptibility graph obtained can better reflect the actual distribution of landslides, which is of great significance for improving the landslide susceptibility in unknown areas.

Under the same graph structure, the number of historical landslide points shows an increasing trend with the increase of the landslide susceptibility levels of the four models, which indicates that all four models have certain positive effects in

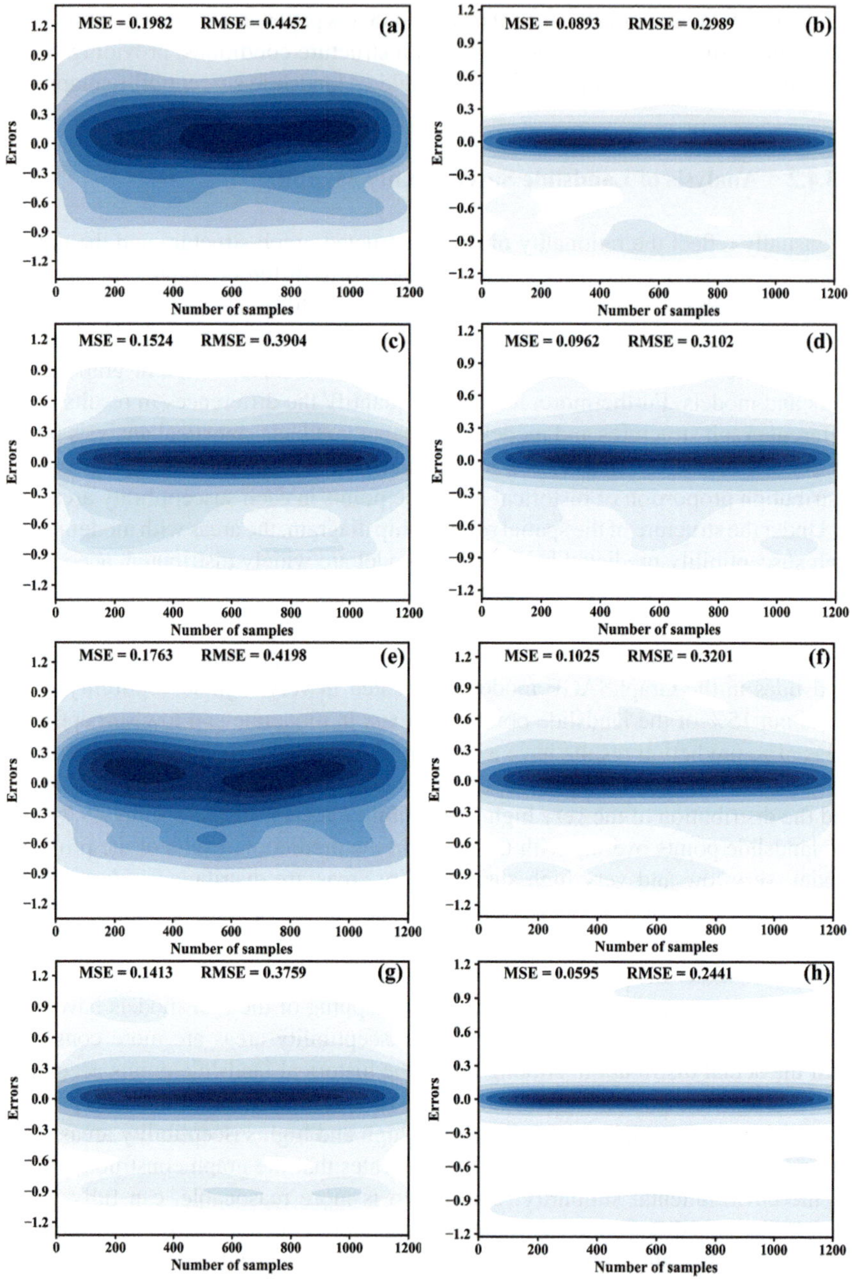

Fig. 4.20 Errors in MSE and RMSE for both the spatial topological relationships graph (left) and environment similarity relationship graph (right): **a**, **b** GCN, **c**, **d** GraphSAGE, **e**, **f** GAT, **g**, **h** Proposed model, reprinted from Zhang et al. (2024b), copyright 2025, with permission from IEEE

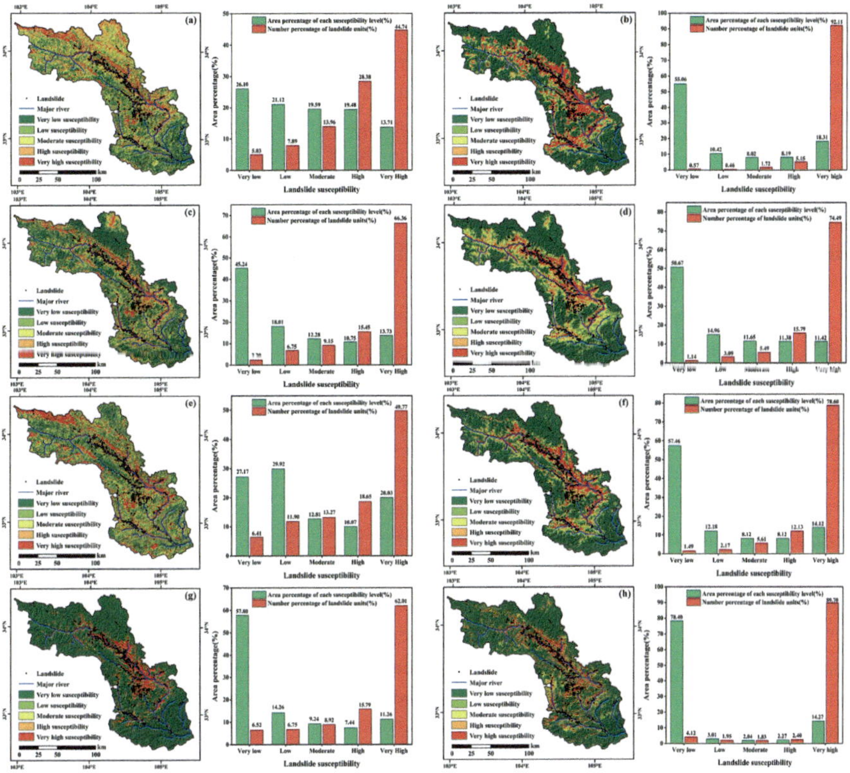

Fig. 4.21 Landslide distribution indifferent graph structures and models for both spatial topological relationships graph (left) and environment similarity relationship graph (right). **a**, **b** GCN. **c**, **d** GraphSAGE. **e**, **f** GAT. **g**, **h** Proposed model, reprinted from Zhang et al. (2024b), copyright 2025, with permission from IEEE

predicting the susceptibility of landslides. However, from the perspective of the distribution matching between historical landslide points and very high susceptibility areas, the model proposed in this paper shows the optimal prediction results of landslide susceptibility. The landslide area predicted by this model is more detailed, and most of the landslide points are concentrated in the very high susceptibility area.

Furthermore, under different graph structures, compared with the landslide susceptibility graphs of the other three models, the landslide susceptibility distribution of the proposed model is more reasonable, and the very high susceptibility areas it divides are in the best agreement with the actual distribution of historical landslides. This result indicates that the model constructed in this paper has excellent robustness and exhibits good performance under two different graph structures. This model can deeply explore the potential relationship between landslides and disaster-prone environments, effectively reduce the false alarm rate and missed alarm rate, and provide valuable reference basis for future landslide risk management.

4.6 Summary

This chapter introduces the landslide susceptibility mapping method based on Graph neural networks under the constraint of geographical environment consistency. By constructing geographical nodes and weighted graph models, there are mainly the GraphSAGE landslide susceptibility mapping model based on environmental information aggregation and the landslide susceptibility mapping model based on Graph-Transformer graph neural networks. The precise mapping of the susceptibility to landslides has been achieved.

Taking the mapping of landslide susceptibility as the core objective, combined with deep learning technology and geographical laws, a landslide susceptibility mapping method based on graph neural networks is proposed. During the research process, the key influencing factors of landslide development were first identified. On this basis, geographical nodes and directed weighted maps were constructed to characterize the spatial correlation characteristics within the region. Secondly, in order to solve the problem of unbalanced landslide distribution, a sampling strategy based on environmental similarity constraints is proposed to ensure the representativeness and balance of the training data. Finally, with the aid of the optimized graph neural network, a landslide susceptibility mapping model capable of integrating environmental information was constructed. The experimental results show that the proposed graph neural network landslide susceptibility mapping method with geographical environment consistency constraints can significantly improve the accuracy and reliability of landslide risk mapping. Through comparative analysis, it is found that this method is superior to the traditional methods in all evaluation indicators, fully demonstrating its superiority and practicability.

Research is carried out around the application of graph neural networks integrating attention mechanisms in the mapping of the susceptibility of wide-area landslides. Firstly, the basic principles of graph neural networks were systematically expounded, emphasizing their unique advantages in capturing spatial dependencies and complex nonlinear features. Then, through in-depth analysis of geospatial data and the characteristics related to landslides, a spatial topological relationship graphc of the landslide area was constructed. Subsequently, the core concept of the attention mechanism was further explored, and the mechanism by which it enhances the feature extraction ability and prediction accuracy of the model through the dynamic allocation of weights was explained. Based on this, a new model named Graph-Transformer is proposed. This model organically combines the structured learning ability of graph neural networks with the global feature capture ability of Transformers, thereby achieving efficient fusion of local and global information. Finally, through multiple sets of comparative experiments, a comprehensive evaluation and detailed analysis of the performance and prediction results of the proposed model were conducted.

References

Bai, S., Wang, J., Zhang, Z., & Cheng, C. (2012). Combined landslide susceptibility mapping after Wenchuan earthquake at the Zhouqu segment in the Bailongjiang Basin, China. *CATENA, 99*, 18–25. https://doi.org/10.1016/j.catena.2012.06.012

Bruna, J., Zaremba, W., Szlam, A., & LeCun, Y. (2014). Spectral networks and locally connected networks on graphs. arXiv. https://doi.org/10.48550/arXiv.1312.6203

Canavesi, V., Segoni, S., Rosi, A., Ting, X., Nery, T., Catani, F., & Casagli, N. (2020). Different approaches to use morphometric attributes in landslide susceptibility mapping based on meso-scale spatial units: A case study in Rio de Janeiro (Brazil). *Remote Sensing, 12*(11), 1826. https://doi.org/10.3390/rs12111826

Catani, F., Lagomarsino, D., Segoni, S., & Tofani, V. (2013). Landslide susceptibility estimation by random forests technique: Sensitivity and scaling issues. *Natural Hazards and Earth System Sciences, 13*(11), 2815–2831. https://doi.org/10.5194/nhess-13-2815-2013

Chang, Z., Huang, J., Huang, F., Bhuyan, K., Meena, S. R., & Catani, F. (2023). Uncertainty analysis of non-landslide sample selection in landslide susceptibility prediction using slope unit-based machine learning models. *Gondwana Research, 117*, 307–320. https://doi.org/10.1016/j.gr.2023.02.007

Dou, J., Yunus, A. P., Merghadi, A., Shirzadi, A., Nguyen, H., Hussain, Y., Avtar, R., Chen, Y., Pham, B. T., & Yamagishi, H. (2020). Different sampling strategies for predicting landslide susceptibilities are deemed less consequential with deep learning. *Science of the Total Environment, 720*, Article 137320. https://doi.org/10.1016/j.scitotenv.2020.137320

Ermini, L., Catani, F., & Casagli, N. (2005). Artificial Neural Networks applied to landslide susceptibility assessment. *Geomorphology, 66*(1–4), 327–343. https://doi.org/10.1016/j.geomorph.2004.09.025

Fu, X., Liu, Y., Zhu, Q., Ge, D., Li, Y., & Zeng, H. (2022). Reliable assessment approach of landslide susceptibility in broad areas based on optimal slope units and negative samples involving priori knowledge. *International Journal of Digital Earth, 15*(12), 2495–2510. https://doi.org/10.1080/17538947.2022.2159549

Guo, C., Montgomery, D. R., Zhang, Y., Wang, K., & Yang, Z. (2015). Quantitative assessment of landslide susceptibility along the Xianshuihe fault zone, Tibetan Plateau, China. *Geomorphology, 248*, 93–110. https://doi.org/10.1016/j.geomorph.2015.07.012

Hakim, W. L., Fadhillah, M. F., Park, S., Pradhan, B., Won, J.-S., & Lee, C.-W. (2023). InSAR time-series analysis and susceptibility mapping for land subsidence in Semarang, Indonesia using convolutional neural network and support vector regression. *Remote Sensing of Environment, 287*, Article 113453. https://doi.org/10.1016/j.rse.2023.113453

Huang, F., Ye, Z., Jiang, S.-H., Huang, J., Chang, Z., & Chen, J. (2021). Uncertainty study of landslide susceptibility prediction considering the different attribute interval numbers of environmental factors and different data-based models. *CATENA, 202*, Article 105250. https://doi.org/10.1016/j.catena.2021.105250

Huang, F., Zhang, J., Zhou, C., Wang, Y., Huang, J., & Zhu, L. (2019). A deep learning algorithm using a fully connected sparse autoencoder neural network for landslide susceptibility prediction. *Landslides, 17*(1), 147–164. https://doi.org/10.1007/s10346-s019-01274-9

Kipf, T. N., & Welling, M. (2017). Semi-supervised classification with graph convolutional networks. In: *Proceedings of the 5th international conference on learning representations (ICLR)*. https://doi.org/10.48550/arXiv.1609.02907

Marjanović, M., Kovačević, M., Bajat, B., & Voženílek, V. (2011). Landslide susceptibility assessment using SVM machine learning algorithm. *Engineering Geology, 123*(3), 225–234. https://doi.org/10.1016/j.enggeo.2011.09.006

Micheli, A. (2009). Neural network for graphs: A contextual constructive approach. *IEEE Transactions on Neural Networks, 20*(3), 498–511. https://doi.org/10.1109/TNN.2008.2010350

O'Brien, R. M. (2007). A caution regarding rules of thumb for variance inflation factors. *Quality & Quantity, 41*(5), 673–690. https://doi.org/10.1007/s11135-006-9018-6

Romstad, B., & Etzelmüller, B. (2012). Mean-curvature watersheds: A simple method for segmentation of a digital elevation model into terrain units. *Geomorphology, 139–140*, 293–302. https://doi.org/10.1016/j.geomorph.2011.10.031

Sameen, M. I., Pradhan, B., & Lee, S. (2020). Application of convolutional neural networks featuring Bayesian optimization for landslide susceptibility assessment. *CATENA, 186*, Article 104249. https://doi.org/10.1016/j.catena.2019.104249

Schlögel, R., Marchesini, I., Alvioli, M., Reichenbach, P., Rossi, M., & Malet, J.-P. (2018). Optimizing landslide susceptibility zonation: Effects of DEM spatial resolution and slope unit delineation on logistic regression models. *Geomorphology, 301*, 10–20. https://doi.org/10.1016/j.geomorph.2017.10.018

Vaswani, A., Shazeer, N., Parmar, N., Uszkoreit, J., Jones, L., Gomez, A. N., Kaiser, L., & Polosukhin, I. (2017). Attention is all you need. *Advances in Neural Information Processing Systems*, 30, 5998–6008. https://doi.org/10.48550/arXiv.1706.03762

Veličković, P., Cucurull, G., Casanova, A., Romero, A., Liò, P., & Bengio, Y. (2018). Graph attention networks. In: *Proceedings of the 6th international conference on learning representations (ICLR)*. https://doi.org/10.48550/arXiv.1710.10903

Wei, R., Ye, C., Sui, T., Ge, Y., Li, Y., & Li, J. (2022). Combining spatial response features and machine learning classifiers for landslide susceptibility mapping. *International Journal of Applied Earth Observation and Geoinformation, 107*, Article 102681. https://doi.org/10.1016/j.jag.2022.102681

Xiao, L., Zhang, Y., & Peng, G. (2018). Landslide susceptibility assessment using integrated deep learning algorithm along the China–Nepal Highway. *Sensors, 18*(12), 4436. https://doi.org/10.3390/s18124436

Yunus, A. P., Fan, X., Tang, X., Jie, D., Xu, Q., & Huang, R. (2020). Decadal vegetation succession from MODIS reveals the spatio-temporal evolution of post-seismic landsliding after the 2008 Wenchuan earthquake. *Remote Sensing of Environment, 236*, Article 111476. https://doi.org/10.1016/j.rse.2019.111476

Zeng, H., Zhu, Q., Ding, Y., Hu, H., Chen, L., Xie, X., Chen, M., & Yao, Y. (2022). Graph neural networks with constraints of environmental consistency for landslide susceptibility evaluation. *International Journal of Geographical Information Science, 36*(11), 2270–2295. https://doi.org/10.1080/13658816.2022.2103819

Zhang, Q., He, Y., Zhang, L., Lu, J., Gao, B., Yang, W., Chen, H., & Zhang, Y. (2024a). A landslide susceptibility assessment method considering the similarity of geographic environments based on graph neural network. *Gondwana Research, 132*, 323–342. https://doi.org/10.1016/j.gr.2024.04.013

Zhang, Q., He, Y., Zhang, Y., Lu, J., Zhang, L., Huo, T., Tang, J., Fang, Y., & Zhang, Y. (2024b). A graph–transformer method for landslide susceptibility mapping. *IEEE Journal of Selected Topics in Applied Earth Observations and Remote Sensing, 17*, 14556–14574. https://doi.org/10.1109/JSTARS.2024.3437751

Zhu, L., Wang, G., Huang, F., Li, Y., Chen, W., & Hong, H. (2022). Landslide susceptibility prediction using sparse feature extraction and machine learning models based on GIS and remote sensing. *IEEE Geoscience and Remote Sensing Letters, 19*, 1–5. https://doi.org/10.1109/LGRS.2021.3054029

Chapter 5
An Integrated Dynamic–Static Feature Framework for Landslide Susceptibility Mapping Using InSAR Based Ground Surface Deformation

The development process of landslides often has time series characteristics such as ground surface deformation. However, most of the existing landslide susceptibility mapping (LSM) methods are based on the static characteristics of influencing factors and have not considered the dynamic characteristics of the landslide occurrence time series, resulting in problems such as poor reliability of the evaluation results of landslide susceptibility. Therefore, in this section, the dynamic and static feature integrated LSM method combining InSAR based ground surface deformation realizes the comprehensive utilization of dynamic and static features of landslides through the parallel deep learning network model, thereby improving the reliability of LSM.

5.1 Research Data

This study takes Lanzhou, Gansu Province, Northwest China as an example. Lanzhou is located in the transitional area from the Qinghai-Tibet Plateau to the Loess Plateau, with large terrain undulations and a complex geological environment. The study area is mainly composed of Quaternary wind-formed loess, with vertical joints and loose structure. It is particularly prone to subsidence during the rainy season. Coupled with the increasingly intense impact of human engineering activities on the geological environment in recent years, landslide disasters occur frequently within the territory. The overview of the study area is described in Sect. 3.1.1.1..

For the research area and content of this paper, three types of data were mainly obtained and constructed, namely: Sentinel-1A SAR image, Sentinel-2A optical remote sensing image and multi-source landslide influencing factors. SAR images are processed by SBAS-InSAR to obtain ground surface deformation information for discovering and explaining the deformation process of landslides. Optical remote sensing images express the optical texture characteristics of landslide occurrence to assist in landslide identification. Multi-source landslide influencing factors represent

Y. He, *InSAR and Deep Learning in Landslides Research: Intelligent Identification, Risk Assessment and Susceptibility Mapping*, https://doi.org/10.1007/978-981-96-9132-6_5

the environmental mechanism of landslide occurrence from different perspectives, used to map the development characteristics of landslides and reveal potentially dangerous areas.

5.1.1 *Time Series InSAR Based Ground Surface Deformation Data*

The specific SBAS-InSAR processing flow for obtaining time series round surface deformation data includes the following six key steps:

(1) Data preprocessing
The obtained Sentinel-1A is appropriately cropped according to the requirements of the study area to improve the processing efficiency and prepare the simultaneous precise orbit ephemerides of the SAR image.
(2) Generation of connection graphs
In order to reduce the influence of incoherence and improve the experimental accuracy. For N + 1 SAR images, baseline estimation is first conducted. Then, based on setting reasonable temporal baseline and spatial baseline thresholds, the main image and secondary image are selected. Interference pairs are formed based on certain temporal and spatial baseline thresholds to generate connection graphs.
(3) Interference process
This step can subtract the elevation error and the low-frequency signal error to obtain the phase unwrapping graph of the residual phase. Firstly, all the images are registered onto the super master image to generate interference fringes. Then, terrain phase removal and filtering are carried out based on the external DEM data and the polynomial model (Goldstein). Finally, the Minimum cost flow (MCF) method is used for untangles. To ensure the accuracy of the experiment, the results of this step are usually screened, and interference pairs with poor unwrapping effect or coherence are removed.
(4) Track refining and re-leveling
During the interference process, the phase of the unwrapping graph has both constant phase and phase jump. Therefore, it can be estimated and removed through orbit refinement and reflattening. This step converts the unwrapping phase into an elevation or deformation value based on the GCP as the reference. In the most ideal situation, the selected point should not have deformation fringes, residual terrain or phase jumps, etc. However, in practice, due to various factors, we select points with good coherence far from the deformation area as control points, such as the low-lying areas of valleys and the locations of buildings like Bridges.
(5) Deformation inversion
This step involves two inversions in total. The first inversion is to estimate the surface deformation rate and the residual terrain, and solve the phase change rate

within each time period by using algorithms such as least squares or singular value decomposition. The second inversion is based on the first inversion. It uses high-pass filtering in the time dimension and low-pass filtering in the space dimension to remove the interference of the atmospheric phase and the noise phase, and obtains the final deformation phase.

(6) Geocoding
This step is to convert the obtained deformation information from the SAR coordinate system to the geographic coordinate system, obtain the line-of-sight deformation rate and the cumulative amount of deformation in the time series. Meanwhile, through calculation, the LOS ground surface deformation can be transformed into the vertical deformation according to the radar incident angle, revealing different variation characteristics.

5.1.2 Landslide Influencing Factors

The landslides influencing factors reflect the complex environmental characteristics of landslide development from different perspectives. Based on existing studies and the actual situation of the study area, this paper selects a total of 10 environmental factors, covering various categories such as topography, geology, hydrology, vegetation coverage, and human engineering activities.

The construction results of the landslide influencing factors are shown in Fig. 5.1. The slope and elevation can provide a certain potential energy for the occurrence of landslides (Pham et al., 2017a). Combined with historical cases, the study area has many mountains, and areas with higher elevations and steep mountain slopes are more prone to landslides. Aspect is related to the degree of influence of precipitation and solar radiation on the region (Chauhan et al., 2010), and affects the occurrence of landslides by influencing water movement, soil properties, etc. Lithology and rock types have a significant influence on the type of slope soil, slope structure and shear strength of soil. The distance-type influencing factor is calculated through the Euclidean distance. The distance from faults affects the mechanical structure of the landslide mass. When the rock and soil mass that constitutes the slope is cut into discontinuous states by various tectonic surfaces, conditions for sliding will occur. The distance to river reflects the fluvial scouring and soil erosion effects of surface runoff on the evolution of landslides. Land use types and road distance reflect human utilization and alteration of natural conditions. Human activities such as building roads and railways, irrigating farmland, and constructing houses along mountains can also increase the frequency and possibility of landslide disasters (Gao et al., 2023). The NDVI is an index used to assess the growth status of surface vegetation. When the value is relatively small, it indicates that the vegetation in the area is growing poorly, the soil stability is poor, and there may be a higher risk of landslides (Sun et al., 2013). Rainfall is an important inducing factor for landslides. Excessive precipitation will increase the water content of surface water and groundwater, change the physical properties of the soil, reduce the shear strength of the soil and increase the soil permeability, thereby intensifying the occurrence of landslides. The detailed data sources and data resolutions are described in Table 5.1.

If the selected landslide influencing factor is regarded as the independent variable and whether there is a landslide or not is regarded as the dependent variable, there exists a complex nonlinear relationship between the two. However, when there is a strong linear relationship among the independent variables, that is, a slight change in one variable cause changing in other variables, it will be very difficult to find the true relationship between multiple independent variables and the dependent variable, which may lead to serious deviations or even completely opposite conclusions. We call such problems multicollinearity problems among variables (O'brien, 2007). The influencing factors of landslides are not completely independent of each other, but whether they are attributed to collinearity requires scientific judgment. Therefore, in this paper, the *VIF* and *T* were selected to conduct collinearity tests on ten influencing factors.

The analysis results are shown in Fig. 5.2. The VIF values of the ten landslide influencing factors selected in this paper are all between 1 and 3. The maximum value occurs at the distance from river (2.18), and the minimum value occurs at the average annual rainfall (1.03). The T value is between 0.4 and 1, with the maximum value being 0.96 and the minimum value being 0.45. The results show that the VIF and T of the selected ten influencing factors all meet the critical value requirements, indicating that they have good independence and there is no collinearity problem. To sum up, the ten influencing factors selected in this paper can all be substituted into the subsequent model learning and result evaluation, which can ensure the validity and accuracy of the data.

After confirming that the selected factors were appropriate, the study wanted to make a preliminary understanding of the influence degree of the influencing factors on landslide. Therefore, the Random Forest (RF) algorithm was selected to rank the feature importance of the selected influencing factors. RF is currently one of the most commonly used methods in fields such as data mining and feature screening. It reduces the variance of the model by combining the prediction results of multiple decision trees, has a strong tolerance for outliers and noise, is capable of handling high-dimensional data, and can effectively analyze nonlinear, collinear and interactive data. And be able to give the variable importance score while analyzing the data (Archer & Kimes, 2008). In this paper, the Out-Of-Bag (OOB) error rate method in RF is selected to calculate the importance scores of different factors relative to the occurrence of landslides. The ranking results of importance from high to low are shown in Figs. 5.6 and 5.3 as follows: Distance to roads (0.17), elevation (0.16), distance to faults (0.15), distance to rivers (0.12), average annual rainfall (0.11), slope (0.08), NDVI (0.05), aspect (0.05), lithology (0.04), land use types (0.01). The distance to roads is the most important. It reveals the serious impact of human engineering activities on the occurrence of landslides. Elevation and faults both provide prerequisites for landslides. Rivers and rainfall promote the occurrence of landslides by affecting soil structure. Other factors are also related to the occurrence of landslides to varying degrees. The overall importance ranking results are reasonable.

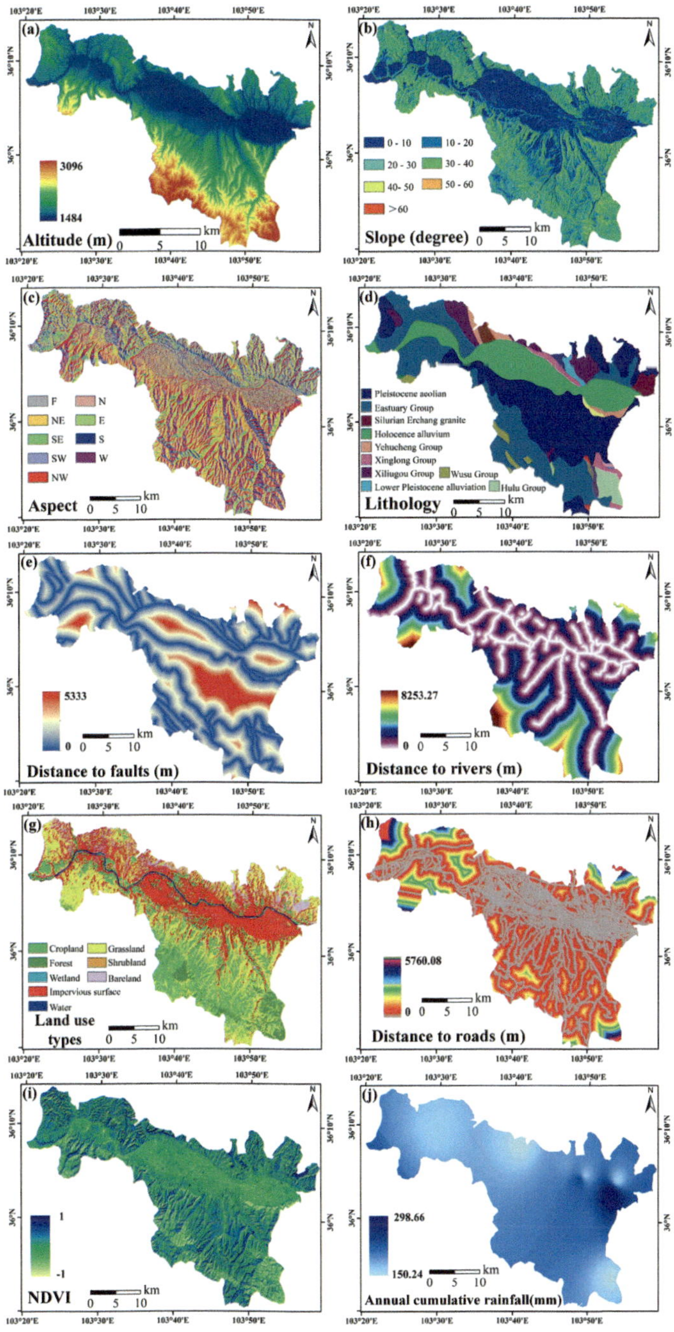

Fig. 5.1 Landslide influencing factors, reprinted from Ref. Zhao et al. (2022), copyright 2025, with permission from Taylor & Francis

Table 5.1 Data sources of landslide influencing factors, reprinted from Ref. Zhao et al. (2022), with permission from Taylor & Francis

Data name	Source	URL	Factors	Resolution
Digital elevation model	United States Geological Survey	earthexplorer.usgs.gov accessed on May 6, 2025	Elevation	30 m
			Slope	
			Aspect	
Land use types	Tsinghua University	http://data.ess.tsinghua.edu.cn/ accessed on May 6, 2025	Land use types	30 m
Rainfall	Geographic remote sensing ecological network	http://www.gisrs.cn/ accessed on May 6, 2025	Rainfall	50 m
NDVI	National Ecological Data Center Resource Sharing Service Platform	http://www.nesdc.org.cn accessed on May 6, 2025	NDVI	10 m
Lithology and fault	United States Geological Survey	https://www.cgs.gov.cn/ accessed on May 6, 2025	Lithology	1:4,000,000
			Distance to faults	30 m
River and road	Department of Resources of Lanzhou	http://zrzyj.lanzhou.gov.cn/ accessed on May 6, 2025	Distance to rivers	30 m
			Distance to roads	30 m

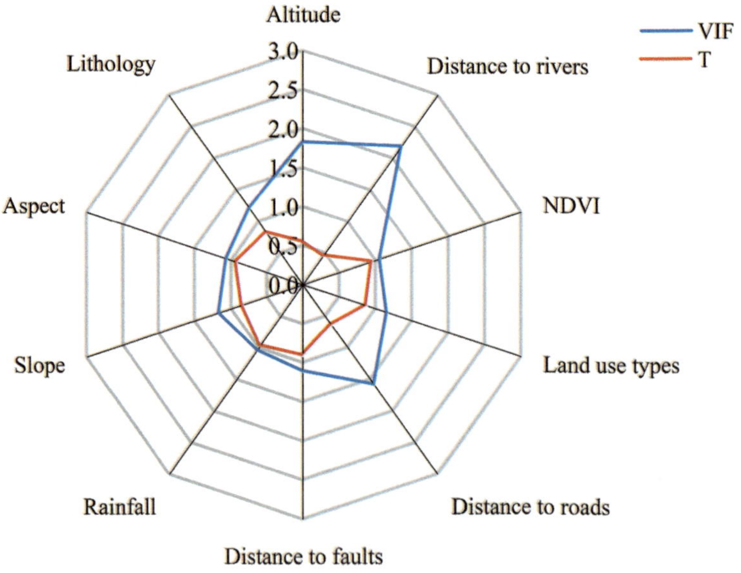

Fig. 5.2 Multicollinearity analysis results of landslide influencing factors, reprinted from Ref. Zhao et al. (2022), copyright 2025, with permission from Taylor & Francis

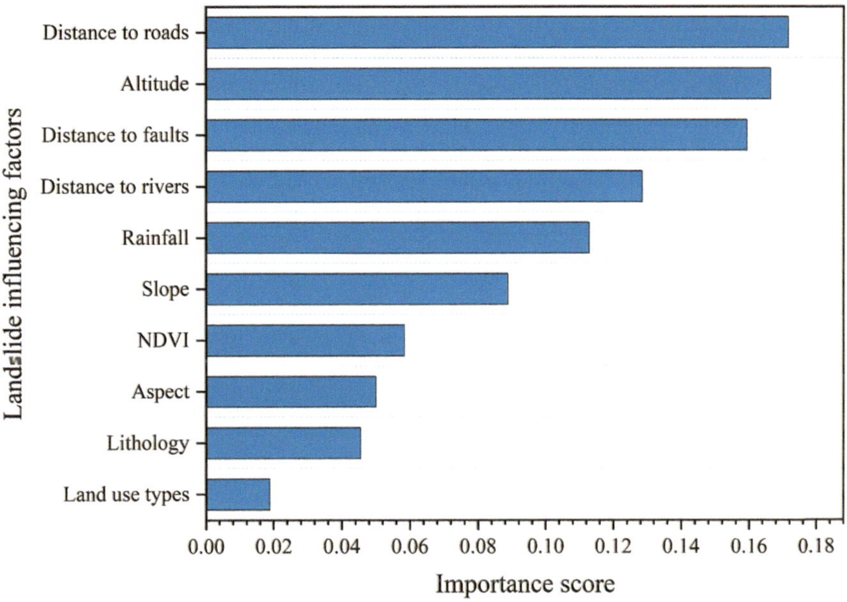

Fig. 5.3 The ranking results of the feature importance of landslide influencing factors, reprinted from Ref. Zhao et al. (2022), copyright 2025, with permission from Taylor & Francis

5.1.3 Landslide Database

The landslide database data used in this chapter originated from the U-Net landslide identification method of the attention mechanism of Sect. 3.1. Optical remote sensing for the landslides in Lanzhou. The multi-band Sentinel-2A optical remote sensing images were introduced for landslide identification, and the U-Net landslide identification model of the channel attention mechanism was constructed, which improved the generalization ability and stability of landslide identification and achieved intelligent and accurate identification of landslides. This method identified a total of 252 landslides, with a total area of 19.16 km^2.

5.2 Landslide Susceptibility Mapping Based on Dynamic and Static Characteristics Combined with InSAR Based Ground Surface Deformation

After obtaining the accurate landslide locations in the study area in the previous text, this chapter, based on the landslide identification results and the historical landslide database, uses deep learning methods to map the landslide susceptibility of the entire study area, aiming to reveal the high-risk areas of landslides in the study area and provide reliable support for disaster prevention and control as well as emergency management. To comprehensively evaluate the development characteristics of landslides and make reasonable and accurate evaluations, this chapter proposes a parallel network model that integrates the dynamic and static characteristics of landslide development. To elaborate on the practicability of this method in the research of landslide susceptibility mapping, it will be fully expounded from aspects such as data processing, model construction, model training and accuracy verification.

5.2.1 Data Processing for Landslide Susceptibility Mapping

5.2.1.1 Data Processing for Landslide Susceptibility Mapping

The most intuitive feature of landslide development lies in the physical displacement of rock and soil masses. In most cases of landslide development, the landslide is often accompanied by a long period of deformation information before the final collapse. Although some landslides have sudden situations, there are also many signs of deformation in the early stage (Gao et al., 2024). However, in the current research on the landslide susceptibility mapping, only start from the perspective of fixed influencing factors and have not considered such developmental characteristics that change over time. There exist problems such as incomplete description of landslide characteristics and unreliable results. Therefore, this paper proposes to use the ground surface deformation observed by InSAR as the feature representing the dynamic information of landslides, thereby obtaining more accurate susceptibility mapping results.

The time series data used in this paper are the ground surface deformation results of the SBAS-InSAR ascending track in the main urban area of Lanzhou from 2015 to 2020, which were completed in the early processing. To meet the input and learning requirements of deep learning models, data preprocessing needs to be carried out first. To preserve the feature distribution and time series relationship of the original data, this paper proposes that dynamic data be processed through two steps: feature interpolation and dynamic normalization.

Feature interpolation is to ensure that the time series data have the same time interval, so as to facilitate subsequent data selection and model exploration of the spatiotemporal development law of landslides. Through the previous SBAS-InSAR processing procedure, the cumulative deformation values of the study area in different periods from January 1, 2015 to June 15, 2020 were obtained as a result. To avoid the influence of incoherence and uncertain factors, the original data chose to obtain images within the smallest possible time interval. Although it was preset to obtain one scene every 12 days, due to factors such as atmospheric interference and flight orbits at some times, 57 SAR images were ultimately obtained, with relatively large intervals between some time periods. To maintain the continuous deformation law, the study proposes to interpolate the data of the missing period without changing the obtained results. Interpolation aims to make the cumulative deformation time intervals of each scene equal, that is, to obtain the ground surface deformation result of one scene every 12 days. The interpolation method is as shown in Eq. (5.1), where D_{t_s} represents the cumulative deformation at the starting time node of the period requiring interpolation and D_{t_e} represents the cumulative deformation at the ending time node of the period requiring interpolation. According to the cumulative deformation differences within the two node time periods, the deformation differences are evenly distributed in each 12-day period and then allocated to each interpolation time. And i represents the i break of the time period that requires interpolation, and D_i is the cumulative deformation result obtained by interpolation for each segment. According to this method, 167 scenes and images were finally interpolated.

$$D_i = D_{t_s} + \frac{12i}{t_s - t_e}\left(D_{t_e} - D_{t_s}\right) \tag{5.1}$$

After obtaining the cumulative deformation of the complete time series, it is necessary to consider how to pass it on to the deep learning model for effective learning. To avoid vanishing or exploding gradients and to ensure learning efficiency, deep learning models usually need to receive data that has been uniformly normalized (Wang et al., 2019). How to normalize the ground surface deformation of each time node while ensuring the characteristics of the time series is the key issue that needs to be considered in this study. The research proposes to take the extreme value of the ground surface deformation as the basis for data normalization, and dynamically stretch the cumulative deformation value within each time node into the relative space. This not only ensures the overall cumulative deformation characteristic distribution but also guarantees that the deformation interval of a single node remains unchanged. The specific calculation is shown as Eq. (5.2). In the formula, X_{p-max} and X_{n-min} respectively represent the maximum positive deformation and the minimum negative deformation in the overall deformation. X_o is the original deformation variable, and X_n is the processed deformation variable. The final data is stretched to the range of 0–255.

$$X_n = 255 \times \frac{X_o - X_{n-\min}}{X_{p-\max} - X_{n-\min}} \tag{5.2}$$

After the above two steps of processing, the InSAR continuous deformation feature completes the supplementation of the time series and meets the data learning requirements of the deep model on the basis of retaining the original feature distribution. Some of the processing results are shown in Fig. 5.4.

5.2.1.2 Histogram Equalization of Influencing Factors and Dataset Construction

The static features used in this paper are the widely used landslide influencing factors at present. Combined with the construction process of the influencing factors in the

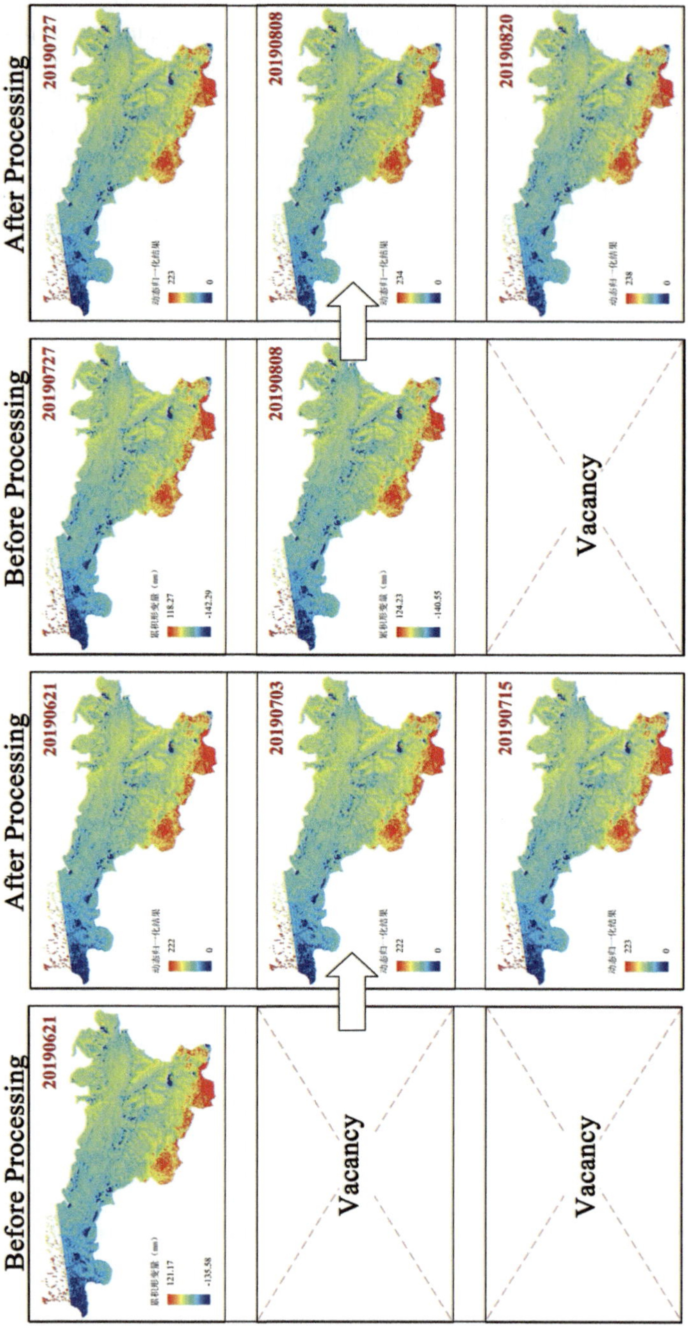

Fig. 5.4 Interpolation and dynamic normalization results of temporal InSAR deformation characteristics

main urban area of Lanzhou, a total of 10 different aspects of influencing factors introduced in the previous text were adopted for the deep learning model to learn. The static influencing factors also need to be normalized before model learning. Due to the different contents represented by each landslide influencing factor and the different expression intervals within the raster, in order to facilitate data analysis and ensure learning efficiency, the study equalized each influencing factor through the histogram. The specific approach is to linearly scale the preprocessed raster of each factor to the range of 0–255 according to their respective extremum values, and then store the single-band bitmap. This method enables each factor image to retain the feature relationship in a single band, ensuring the learning accuracy and accelerating the data reading and processing speed. Meanwhile, a unified linear normalization scheme can be used before entering the model. It is convenient for subsequent model training and evaluation.

In terms of dataset construction, the experiment finally selected a total of 24 periods of ascending track cumulative deformation with 10 influencing factors as static features and one scene taken every 96 days of each quarter as dynamic features to form the overall dataset, and the center point of the landslide occurrence was used as the basis for sample extraction. To summarize the landslide risks in the study area, the research utilized the landslide locations independently identified in the early stage and the historical landslide locations that have been recorded. In this paper, there are a total of 160 autonomous landslide identification points. Historical landslide points include 521 loess landslide disaster points and 216 rain-type landslides. Due to some overlaps, a total of 832 landslides were finally obtained in the study area. During the construction of the dataset, in order to enable the model to fully learn the features, data augmentation processing is introduced for the landslide samples. The limited positive landslide samples and the corresponding factor images are expanded to the dataset through horizontal and vertical flipping. After data augmentation processing, 2,667 landslide samples were collected and 3,000 non-landslide points were randomly selected. Ultimately, the landslide and non-landslide samples constitute the overall data set. After shuffling the order of the dataset, it was originally divided into the training set, the validation set and the test set in an 8:2:1 ratio. The training dataset is used for the model to learn the landslide characteristics and train the internal parameters of the model. The validation dataset is used to verify the training status of the model and serves as a reference for optimizing the hyperparameters of the model. The test dataset is used to examine the final generalization error of the model and evaluate the robustness of the model. The overall process is shown in Fig. 5.5.

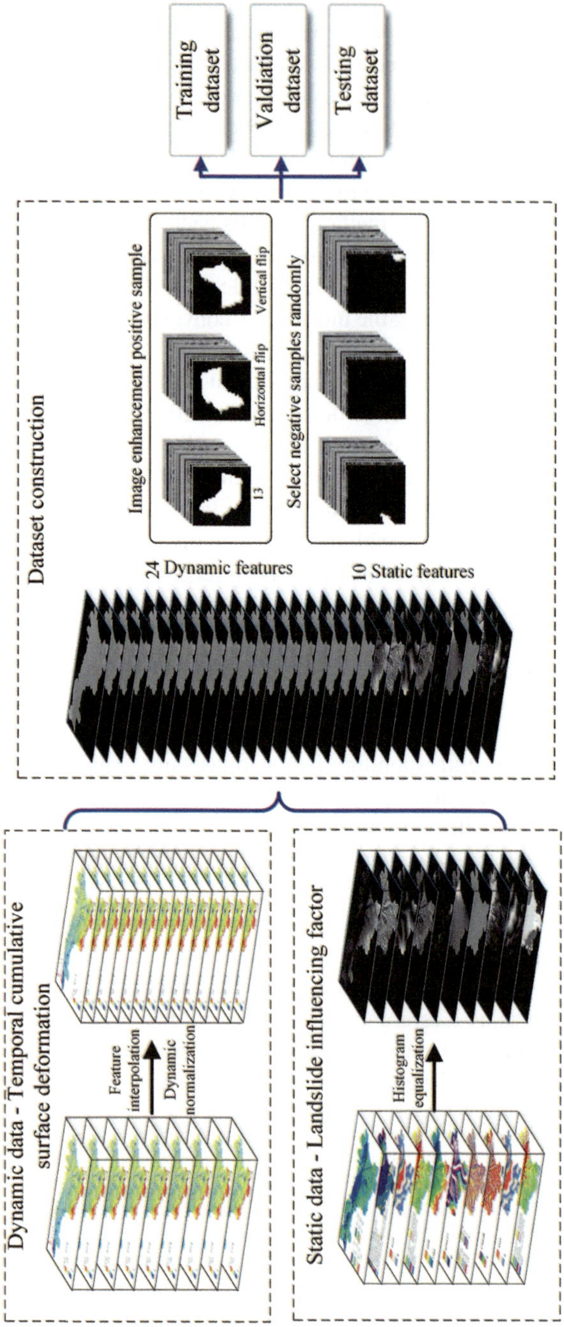

Fig. 5.5 Dataset construction process

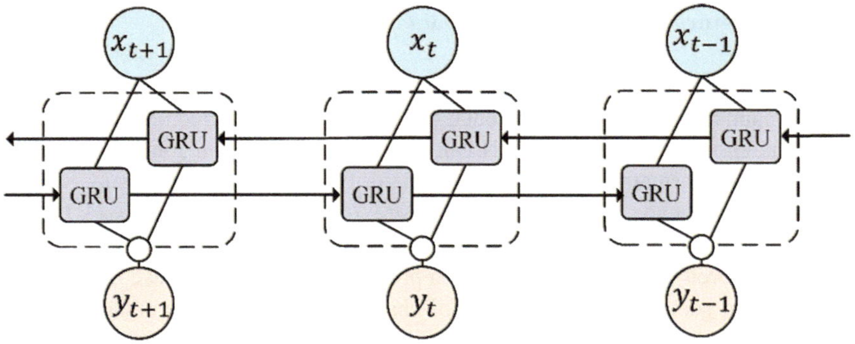

Fig. 5.6 Bi-GRU structure, reprinted from Ref. Zhao et al. (2022), copyright 2025, with permission from Taylor & Francis

5.2.2 *Construction of an Integrated Model for Dynamic and Static Characteristics of Landslides*

5.2.2.1 Feature Learning of Dynamic Data

RNN is a deep learning model often adopted for temporal feature learning, while GRU and Long-Short Term Memory (LSTM) networks are two types of optimized variants of RNN, avoiding the defects of gradient vanishing and gradient explosion of traditional RNN. LSTM uses three gate functions, namely input gate, forget gate and output gate, to control the input value, forget value and output value. The GRU structure simplifies LSTM by combining the forget gate and the input gate into an update gate. A single gating unit is used to simultaneously determine the forget factor and the update state. The network is composed only of two gate functions: the update gate and the reset gate (Zhao et al., 2022).

The single-directional GRU model only receives the sequence feature information from front to back in the input order and can only model use the information before the current time step. Therefore, it may ignore some important information in the time series data, resulting in a decrease in the expressive ability of the model. Accordingly, the proposal of the Bidirectional Gated Recurrent Unit (Bi-GRU) has well solved such problems. Bi-GRU can utilize the forward and backward information in time series data for modeling, thereby improving the expressiveness and robustness of the model. In the Bi-GRU network structure, the hidden structure is composed of two layers of GRU for the forward input and the reverse input superimposed, and its output is jointly determined by the states of the two GRU units at the current moment (Fig. 5.6). This structure, while retaining the structural characteristics of the GRU unit, improves the limitation of the single input sequence, pays more attention to the correlation of time series data before and after, and ensures the full learning of time series features. Therefore, the study intends to adopt Bi-GRU for learning the dynamic temporal characteristics of landslides.

5.2.2.2 Feature Learning of Static Data

The static data of landslides is composed of numerous factors that induce the occurrence of landslides. For such static data, feature learning is carried out by extracting spatial neighborhood features. CNN can actively learn image features and is good at extracting information in the spatial domain. However, the traditional CNN is a feedforward sequence neural network structure. By default, the last network layer is taken as the output, without considering the characteristics of semantic information at different levels in the feature extraction process. It is difficult to take into account the low-level semantic characteristics of shallow features. At the same time, the spatial scale of the feature map changes due to pooling, resulting in information loss. Therefore, this study proposes to extract two-dimensional static features from the landslide static factor dataset using the Multi-scale Convolutional Neural Network (MSCNN), construct a multi-deep fusion feature map, and establish feature learning processes at different scales.

Based on the convolution and pooling layers of traditional CNN, MSCNN uses multi-scale techniques to add fully connected layers after each pooling layer to extract shallow features. The network structure is shown in Fig. 5.7. The *flattening* layer is used to compress multi-dimensional features into one-dimensional vectors. The output features of the middle layer and the last layer are flattened and connected. The feature results extracted by the three fully connected layers are then fused and activated through the *concat* layer to establish a learning process that takes into account both shallow and deep features simultaneously. The specific operation process is shown as follows:

$$h^i = f\left(W^i x^i + b^i\right) \tag{5.3}$$

$$h = \sum_{i=1}^{n} h^i \tag{5.4}$$

h^i ($i = 1,2,3,...n$; n represents the total number of fully connected layers) indicates the i-th fully connected layer, x^i represents the expanded feature of the i-th layer, W^i and b^i respectively represent the corresponding weight matrix and bias, and $f(\cdot)$ is the construction function of the fully connected layer. The fusion and addition of all fully connected layers obtain h, which is the multi-scale feature fusion result.

5.2.2.3 Integrated Model of Dynamic and Static Characteristics of Landslides

Based on the learning advantages of Bi-GRU and MSCNN in terms of time series and space, a landslide susceptibility mapping method is proposed by using the above model to learn deep features from two types of dynamic and static data, namely

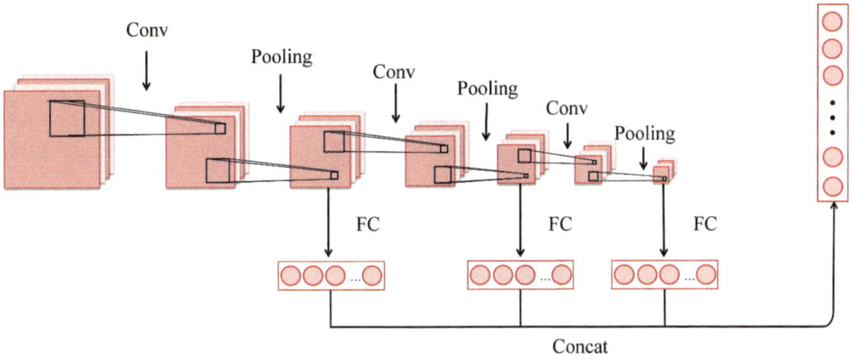

Fig. 5.7 Network structure of MSCNN, reprinted from Ref. Zhao et al. (2022), copyright 2025, with permission from Taylor & Francis

temporal InSAR deformation information and landslide factors, and make comprehensive use of them. The specific model construction idea is shown in Fig. 5.8. The entire network conducts parallel learning of dynamic and static features. After completing the learning, the two types of features are fused to provide the level of landslide susceptibility.

Under the spatial–temporal data organization form that needs to take into account the spatial dimension, the Bi-GRU network has insufficient applicability in expressing the spatial neighborhood characteristics of the data and it is difficult to construct accurate spatial characteristics. Therefore, before the Bi-GRU accepts the data, this structure, the Time Distributed Convolutional Neural Network (TD-CNN) is added to learn and retain the spatial features of the time series. The time distribution approach mainly relies on the Time Distributed wrapping layer under *Keras*. It can apply a common layer to each time step of the time series data and combine the output of each time step into a tensor as the output of this layer (Luna-Alvarez et al., 2020). By applying this technology, the layers applicable to a single time step can be extended to the entire time series, thereby achieving the modeling of time series data. After extracting spatial–temporal feature maps with a smaller spatial size but a higher degree of semantic abstraction through multiple temporal distribution convolution and pooling, the spatial–temporal feature maps are sampled as temporal feature vectors using global pooling. The *Concat* operation is used to merge the feature vectors output by global average pooling and global maximum pooling to reduce the loss of spatial semantic information caused by the global pooling operation. The temporal feature vectors extracted through the above process are passed to the multi-layer Bi-GRU for temporal feature extraction. The last layer of Bi-GRU outputs the feature vector result. The static features of the influence factor were learned by MSCNN in the previous section. For feature maps of different depths and sizes, two convolution operations were conducted respectively to enhance feature learning. As the network depth deepened layer by layer, the final combined deep and shallow feature learning vectors were concatenated and merged with the dynamic

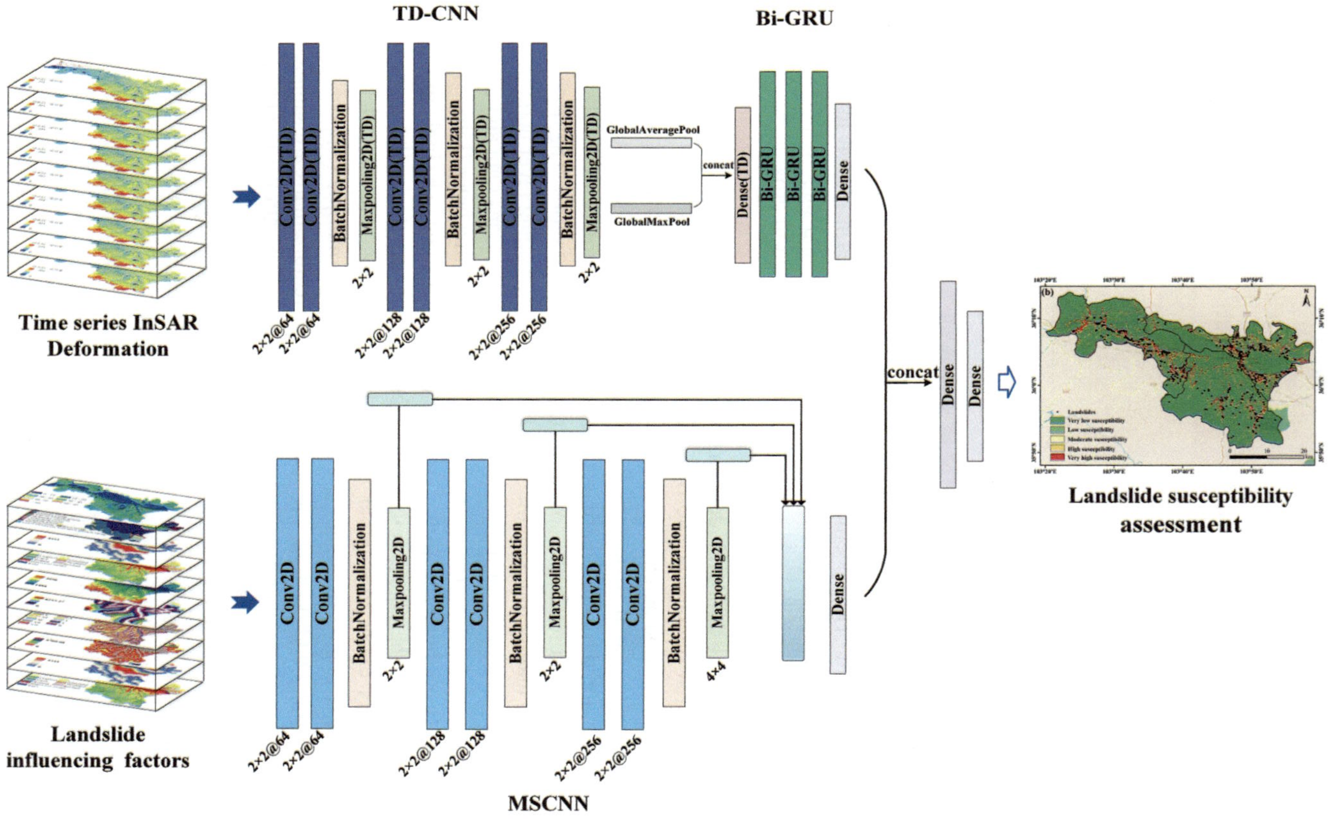

Fig. 5.8 Structure of the integrated model of dynamic and static features, reprinted from Ref. Zhao et al. (2022), copyright 2025,with permission from Taylor & Francis

feature learning in the channel dimension as well. Then, the fused results are learned by the fully connected layer and output as the basis for the final susceptibility of landslides.

We made multiple adjustments to the model parameters using the trial-and-error method, and then obtained the final result. The specific parameter design of the model is presented in Table 5.2. Most of the network layer activation functions in the model are expressed using Rectified Linear units (ReLU), aiming to avoid the insufficient response ability of simple linear relationships among complex variables, and to approximately fit the output results with the real data using nonlinear functions. Meanwhile, the ReLU function is one of the most common and effective activation functions in deep learning. It can help overcome the problem of vanishing gradients and has high efficiency in training, which can significantly shorten the learning cycle of the network. In the output layer, all four models adopt the sigmoid function for output, mapping the output results within the distribution interval from 0 to 1. It can be known in Table 5.2 that after performing two convolution operations on each feature map, we added BN layers to accelerate the training. Meanwhile, the dropout operation was appropriately added in the network hierarchy. Due to space limitations, it is not reflected in the table. The dropout operation can temporarily discard neural network units according to specific probabilities during the training process to reduce the co-adaptation between overfitting and hidden units, and at the same time enhance the generalization ability of the prediction method. From the perspective of network optimization, Adam Optimizer was used instead of the traditional stochastic gradient descent method to update the network weights more effectively. The learning rate was dynamically adjusted through the callback function. The initial learning rate of the network was set to 0.001. During the iteration process, as the loss function of the test set was observed, when every three iterations with equal performance occurred, the learning rate decays at a rate of 0.5 times.

For the output of landslide susceptibility mapping, it is necessary to obtain the probability of landslide occurrence at each point. Therefore, such problems should be regarded as regression problems, and choosing an appropriate loss function is particularly important for the presentation of the results. The MSE is a commonly used loss function in regression problems and a relatively convenient method for measuring the average error. In this study, MSE was selected as the loss function. MSE can evaluate the degree of change of the data. The smaller the value of MSE is, the better the accuracy of the prediction model in describing the experimental data. The calculation of MSE is shown in Formula (5.5). In the formula, $\hat{y}_i$ is the model result, y_i is the real label, and n is the number of samples.

$$\mathrm{MSE} = \frac{1}{\mathrm{n}} \sum_{i=1}^{n} (y_i - \hat{y}_i)^2 \tag{5.5}$$

Table 5.2 Specific parameters of the integrated model of dynamic and static features

Layer function	Layer name	Feature map size	Parameter setting	Activation function
Dynamic data input	Input_layer1	H × W × T × N	–	–
Dynamic feature learning	Conv2D(TD)	H × W × T × C	2 × 2 × 64	Relu
	Conv2D(TD)	H × W × T × C	2 × 2 × 64	Relu
	BatchNormalization	–	–	–
	Maxpooling2D(TD)	H/2 × W/2 × T × C	2 × 2	–
	Conv2D(TD)	H/2 × W/2 × T × 2C	2 × 2 × 128	Relu
	Conv2D(TD)	H/2 × W/2 × T × 2C	2 × 2 × 128	Relu
	BatchNormalization	–	–	–
	Maxpooling2D(TD)	H/4 × W/4 × T × 2C	2 × 2	–
	Conv2D(TD)	H/4 × W/4 × T × 4C	2 × 2 × 256	Relu
	Conv2D(TD)	H/4 × W/4 × T × 4C	2 × 2 × 256	Relu
	BatchNormalization	–	–	–
	GlobelMaxpooling2D(TD)	T × 4C	–	–
	GlobelAvereagepooling2D(TD)	T × 4C	–	–
	Concatenate	T × 8C	–	–
	Dense(TD)	T × 4C	–	Relu
	Bi-GRU	T × 4C	256	tanh
	Bi-GRU	T × 4C	256	tanh
	Bi-GRU	4C	256	tanh
	Dense	2C	128	Relu
Static data input	Input_layer2	H × W × N	–	–
Static feature learning	Conv2D	H × W × C	2 × 2 × 64	Relu
	Conv2D	H × W × C	2 × 2 × 64	Relu
	BatchNormalization	–	–	–
	Maxpooling2D	H/2 × W/2 × C	2 × 2	–
	Flatten + Dense	2C	128	Relu
	Conv2D	H/2 × W/2 × 2C	2 × 2 × 128	Relu

(continued)

Table 5.2 (continued)

Layer function	Layer name	Feature map size	Parameter setting	Activation function
	Conv2D	H/2 × W/2 × 2C	2 × 2 × 128	Relu
	BatchNormalization	–	–	–
	Maxpooling2D	H/4 × W/4 × 2C	2 × 2	–
	Flatten + Dense	2C	128	Relu
	Conv2D	H/4 × W/4 × 4C	2 × 2 × 256	Relu
	Conv2D	H/4 × W/4 × 4C	2 × 2 × 256	Relu
	BatchNormalization	–	–	–
	Maxpooling2D	H/8 × W/8 × 4C	4 × 4	–
	Flatten + Dense	2C	128	Relu
	Concatenate	6C	–	–
	Dense	2C	128	Relu
Comprehensive feature learning	Concatenate	4C	–	–
	Dense	2C	128	Relu
	Dense	1	1	Sigmoid

5.3 Evaluation and Analysis of Landslide Susceptibility Mapping

5.3.1 Comparative Study on Model Performance

To verify the effect and credibility of the integrated evaluation model of landslide susceptibility based on the combined dynamic and static features proposed in this paper, a comparative study is established on the basis of conducting experiments. The comparison target is the landslide susceptibility mapping method using only static data, and the specific structure is the MSCNN network structure indicated in the paper for learning the characteristics of static data. For the control variables, except for the difference in whether dynamic feature learning is included, the rest of the data input forms, hyperparameters, etc. of the two models remain consistent. Comparative studies are conducted on the data of the validation set and the test set respectively.

The changes in the MSE loss value and Accuracy (ACC) of the validation set during the training process reveal the learning effect and generalization ability of the model. Therefore, this paper first uses the training process curve of the validation dataset for comparative analysis, and the training results are shown in Fig. 5.9. It can be seen from the training curves of the two models that the gradual decrease of the MSE value and the gradual increase of the ACC value reflect the learning ability of the models. During this process, both methods adopted the error back-propagation method of the gradient descent algorithm as the training rule. During the gradient descent iteration process, the weights between neurons were adjusted according to the error between the actual value and the predicted value of the training samples, thereby gradually calculating the minimum value of the target loss function and the optimal weight of the model. The curves that eventually tend to be stable express that both models have formed an ordered, stable and decision-making structure. During the training process, the MSCNN structure using only static data has significant oscillations in the initial number of iterations, indicating that the model is sensitive to abnormal data performance in the initial stage and has a slight over-fitting phenomenon in the features received in the early stage. However, the two values return to the normal trajectory in the later stage, indicating that the model has robust learning ability. In contrast, both MSE and ACC using the dynamic and static joint feature model are in a normal changing trend, indicating that the model is not disturbed by the accepted complex features. The model has a relatively stable learning state for static and dynamic features and gradually enhances. However, compared with MSCNN, the model undergoes a slightly larger number of iterations before reaching stability. It indicates that the model parameters need to be adjusted according to complex features, which requires a certain number of learning sessions. From the perspective of training results, the comprehensive utilization of the feature model achieved more accurate training results compared to only using the static feature model. The ACC value reached 0.82, which was 2.7% higher than that of MSCNN (0.79). The MSE value was 0.06, which was 1.1% lower than that of MSCNN (0.07). It indicates that the comprehensive utilization of the dynamic and static feature model can better grasp the landslide characteristics compared with the static data model only, and the result performance is better, which proves the correctness of the construction method in this paper.

The test set data is relatively unfamiliar to the model, which can more objectively demonstrate the learning effect of the model and its generalization ability. The test

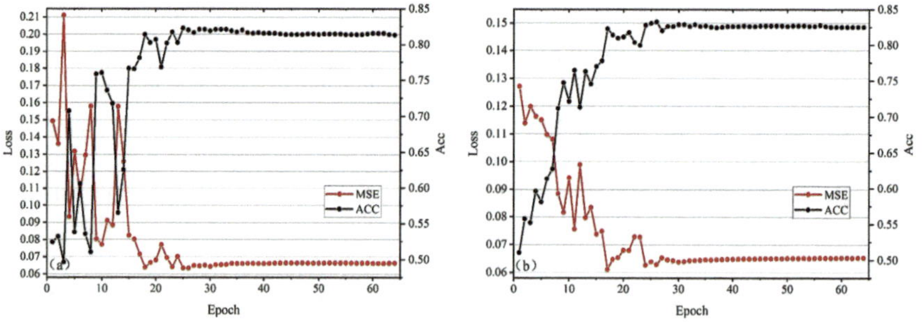

Fig. 5.9 Comparison of model training results, reprinted from Ref. Zhao et al. (2022), copyright 2025, with permission from Taylor & Francis

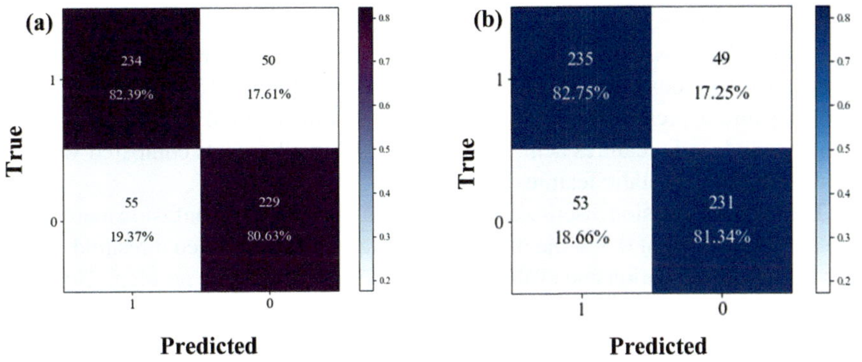

Fig. 5.10 Comparison of the confusion matrix results of the test set, **a** MSCNN, **b** the proposed model, reprinted from Ref. Zhao et al. (2022), copyright 2025,with permission from Taylor & Francis

results of the test set are presented in the form of a confusion matrix to show the comparison between the predicted results and the true values. Since the output result of the model is a probability distribution ranging from 0 to 1, the prediction results are divided into slippery and non-slippery with a threshold of 0.5. The division results of the confusion matrices obtained by the two models are shown in Fig. 5.10.

In the confusion matrix, both models have high true positive rate (TPR) and true negative rate (TNR) in the test set, indicating the reliable classification of the two methods when facing unfamiliar data. In Fig. 5.10, (b) indicates that the proposed method has achieved good classification results for both landslide (1) and non-landslide data (0) in the test set, further demonstrating the model's learning ability. It is worth noting that the issue of landslide susceptibility mapping is not a simple binary classification problem. Therefore, the classification situations of different categories should be analysed in combination with the actual situation. On the one hand, for the landslide data with a true value of 1, we expect the susceptibility mapping model to learn and classify accurately. Only when the susceptibility mapping model truly learns the landslide characteristics can it make correct indications for the risks of landslides that have not occurred. Combining the results of the confusion matrices of the two types of models, the classification effect for the true value of 1 is superior to that for the true value of 0, which proves the learning ability for landslide features. Moreover, the method in this paper performs better, indicating that the learning of landslide features is more thorough. On the other hand, the classification effect of non-landslide data with a true value of 0 should be analysed specifically, as the susceptibility mapping model may classify high-risk areas where landslides have not yet occurred as landslide types. From the results of the confusion matrix, it can be seen that both types of models have some classification errors, which may reveal certain areas with a high incidence of landslides in the study area. There are slightly more errors in the division using only static data. However, the accurate evaluation of the susceptibility to landslides should be based on the accurate learning of landslide characteristics. The method proposed in this paper performs well in landslide samples and has relatively reasonable classification results in the classification of non-landslide areas. From a more objective perspective, the study calculated each corresponding index based on the obtained confusion Table 5.3 matrix. The calculation structure is presented in. The results show that the method using the combined dynamic and static features achieves better results in all aspects compared with the method using only static features.

The above discussion discusses the situation of generating a confusion matrix with the midpoint value of 0.5 of the model's predicted values as a fixed threshold, which is a typical classification and evaluation method for the landslide problem. However, the problem of evaluating the susceptibility of landslides is not a simple binary classification problem. When the thresholds are different, the true positive rate (TPR) and false positive rate (FPR) within the matrix may change. By matching the ratios under the variable thresholds one by one, the receiver operating characteristic curve (ROC) can be plotted. ROC is a synthesis of the confusion matrices under different thresholds, reflecting the continuous changes of data specificity and sensitivity, and is widely used in the evaluation research of landslide susceptibility (Jiang et al., 2024; Pham et al., 2017b). The AUC can visually reflect the results. The closer the AUC value is to 1, the better the model effect is. The ROC performance of the two models in the test set was calculated in the study, and the results are shown in Fig. 5.10. It can be concluded from this that the AUC of the method proposed in this paper reaches 0.90, which is superior to the learning results of the model that only utilizes static features.

Based on the above indicators, the method proposed in this paper for comprehensively utilizing the dynamic and static characteristics of landslides performs outstandingly in all aspects compared with the traditional method that only utilizes static data. The model constructed in this paper can not only learn the complex dynamic and static characteristics of landslides, but also has a stable learning process and high robustness. Meanwhile, the performance in the test set proves the generalization ability. We believe that this model shows applicability in the research on the evaluation of landslide susceptibility and can obtain more accurate and reliable results.

Table 5.3 Comparison of evaluation indicators

Evaluation index	Only utilize static features	Combine dynamic and static features
Accuracy	0.792	0.820
Precision	0.794	0.816
Recall	0.788	0.827
Specificity	0.795	0.813
F1-score	0.791	0.821
Kappa	0.584	0.640

5.3.2 *Analysis of Landslide Susceptibility Mapping*

Reassigning values to each pixel point of the study area using the trained model can produce a landslide susceptibility map of the entire study area. After the model generates the results, the obtained landslide probabilities are classified into five landslide susceptibility grades by the natural breakpoint method using ArcGIS: very low, low, moderate, high, and very high. The RUC and mapping result is shown in Figs. 5.11 and 5.12. Among them, (a) is the susceptibility evaluation result obtained only by using static data; (b) It is the integrated evaluation result of landslide susceptibility proposed in this paper based on the combined dynamic and static characteristics.

As can be seen from Fig. 5.12, there are differences between the results of the two models, but there are still certain commonalities. The prediction locations of the two methods for the very high areas are roughly the same. The vast majority of historical landslide points are included in the very high areas evaluated by the model, which proves the reliability of the results. The predicted results of the two methods show significant differentiation towards the very low susceptibility level areas. More than half of the areas are identified as very low susceptibility. The very high susceptibility areas are distributed along the mountain orientation in the northern and southern mountainous areas, which is consistent with the actual situation of the study area. The method proposed in this paper has relatively fewer very high-risk areas and more precise location characterization.

To quantify the statistical results of landslide susceptibility assessment, we compare the historical landslide distribution with the overall evaluation results.

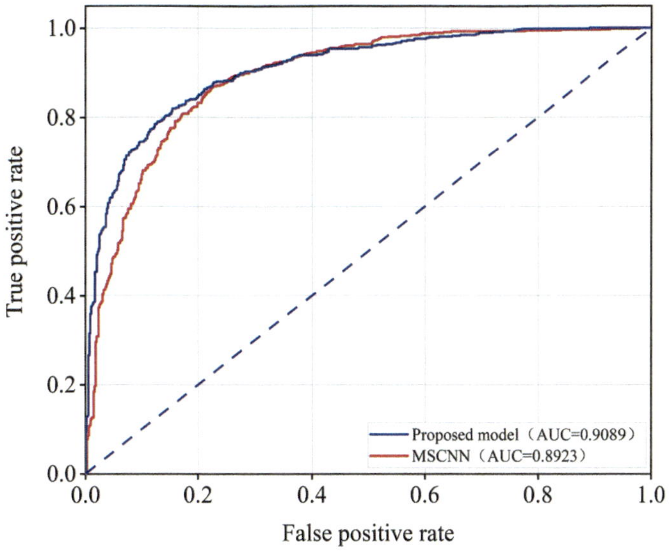

Fig. 5.11 ROC comparison, reprinted from Ref. Zhao et al. (2022), copyright 2025, with permission from Taylor & Francis

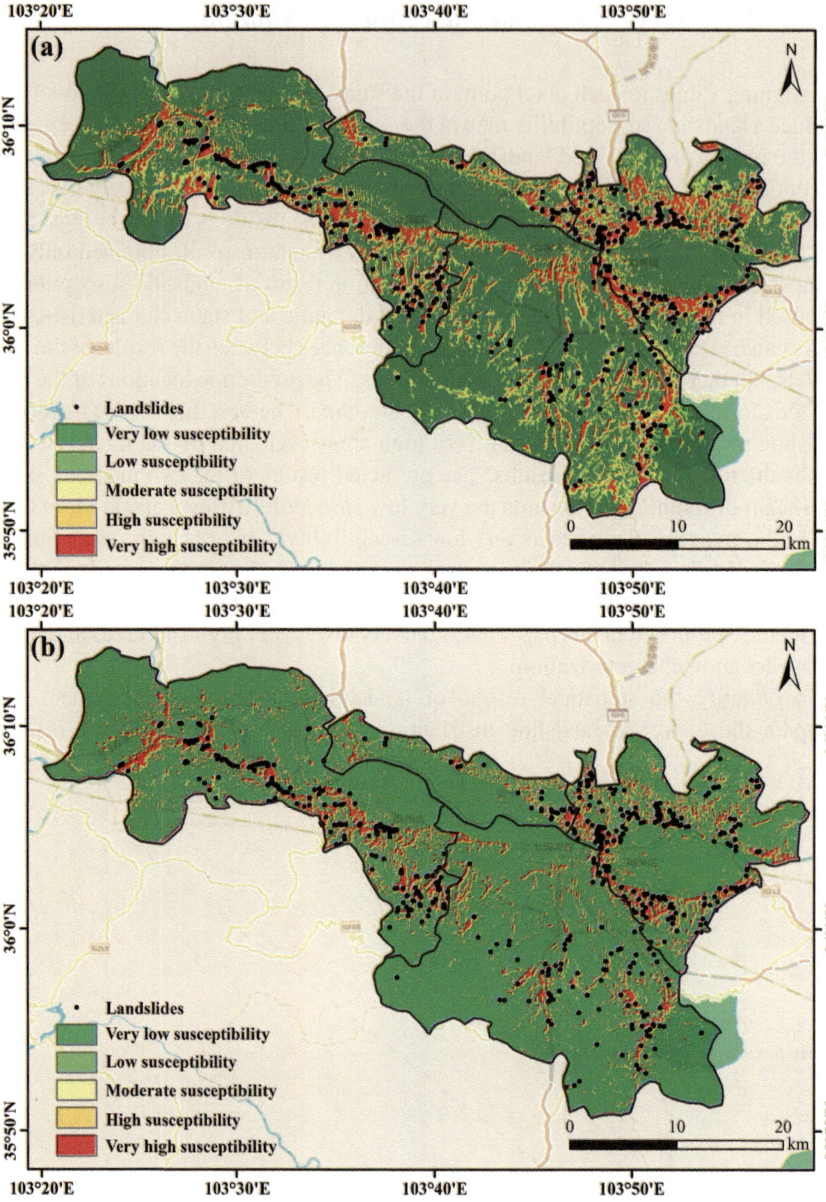

Fig. 5.12 Landslide susceptibility mapping, **a** MSCNN, **b** the proposed model, reprinted from Ref. Zhao et al. (2022), copyright 2025, with permission from Taylor & Francis

The results were objectively evaluated by the ratio of the historical landslide area distributed in each susceptibility grade sub-zone to the area of each susceptibility grade sub-zone. The results are shown in Table 5.4, where Model a and Model b respectively represent the evaluation results of landslide susceptibility obtained by using only static features and the dynamic-static combined feature model. Both the high and very high susceptibility areas of the two models contain more than 80% of historical landslides, and the ratio of historical landslides in different zones of each model shows an upward trend with the increase of susceptibility level, indicating that the models can effectively map the susceptibility of landslides in the urban area of Lanzhou. However, the results of the static feature model are relatively divergent. The distribution ratios of historical landslides at moderate, high and very high levels all exceed 1, and there is a certain degree of landslide misjudgment. In contrast, the proposed method has more reliable results. It has the largest ratio in very high-risk areas and accurately covers 84.79% of the historical landslide range within only 10.18% of the study area, demonstrating a strong feature learning ability. Therefore, this paper holds that constructing a model has higher accuracy compared to using only static data models, and the mapping results of landslide susceptibility in the study area obtained have important reference value.

After obtaining the reliable mapping results of landslide susceptibility mapping, in order to further understand the development law of landslides in the study area and provide clear suggestions for landslide prevention and control, this paper uses the Frequency Ratio (FR) method to conduct a simultaneous analysis of the very high-risk areas and environmental factors in the mapping results, and quantitatively analyze the multiple possible mechanisms in the high-risk areas. It is convenient to adapt measures to local conditions (Achu et al., 2023). The FR calculation is shown in Eq. (5.6), where A represents the number of landslide grids for each type of environmental factor within the interval, A' represents the total number of all landslide grids in the study area, B is the number of grids for the environmental factor within the interval, and B' is the total number of grids in the entire study area. The higher the FR value, the greater the impact of this type of attribute factor on landslides.

Table 5.4 Test table of frequency ratio of landslide susceptibility zones

Susceptibility level	Area proportion	The proportion of historical landslide	Ratio			
	Model a	Model b	Model a	Model b	Model a	Model b
Very low	71.12	78.21	10.92	5.81	0.15	0.07
Low	5.84	5.06	4.88	2.43	0.84	0.48
Moderate	3.78	3.21	4.77	2.56	1.26	0.80
High	4.2	3.24	7.2	4.41	1.71	1.36
Very high	15.07	10.18	72.24	84.79	4.79	8.33

$$FR = \frac{A/A'}{B/B'} \tag{5.6}$$

It can be known from the frequency ratio results (Fig. 5.13) that since more than half of the areas in the main urban area of Lanzhou have an altitude of less than 2000 m, most of the areas with very high landslide occurrence rates are also distributed here. The slope provides the necessary open space conditions for the occurrence of landslides. With the increase of the slope, its influence on the occurrence of landslides gradually increases. The maximum FR value (1.94) of the slope in the study area is manifested between 40 and 50°. The areas prone to landslides are distributed in all slope directions without obvious patterns. Among them, the middle direction has the highest FR value (1.27). Among the land use types, grassland is the area most prone to landslides. The FR value of forest land (0.17) is the lowest, indicating that high vegetation coverage has a certain inhibitory effect on the occurrence of landslides. It is worth noting that the FR value of impermeable water surface (1.10) ranks second, indicating that urban construction activities have a significant impact on the occurrence of landslides. The middle and lower sedimentary ice of the lithology has the greatest impact on the occurrence of landslides, and this lithology is mainly distributed in the southern mountainous area of Chengguan District. The distances from roads, faults and water systems all show a negative correlation with landslides. The farther the distance, the smaller the FR value and the less contribution to landslides. This proves that the development of landslides is indeed affected by human activities, active faults and soil erosion. The overall rainfall in Lanzhou City is relatively low. More than 80% of the areas have an average annual rainfall of less than 20 mm, and the areas with an annual rainfall of 220–240 mm have the highest FR value (2.00). The NDVI value further indicates that most landslides are distributed in areas with poor vegetation growth, and the FR value (1.08) is the highest between −0.5 and 1.

Based on the above analysis of the very high susceptibility areas of landslides and environmental factors, the landslides in Lanzhou are mainly distributed in areas with complex geological structures, high soil moisture content, and frequent human engineering activities. From the perspective of landslide disaster prevention and control, this paper puts forward the following suggestions to avoid the losses of landslide:

(1) Increase vegetation coverage: Appropriately increasing vegetation coverage can effectively reduce soil erosion and surface scouring, and enhance soil stability and anti-slip properties. In view of the current situation in Lanzhou, the government can increase the vegetation coverage level through afforestation, lawn construction, and the promotion of ecological projects, etc., to promote the improvement of the geological environment and ecological protection, and reduce the occurrence of geological disasters.
(2) Control human engineering activities: Excessive human engineering activities can damage the integrity and stability of the land and accelerate the occurrence of landslides. Due to the special terrain in Lanzhou, the relationship between people

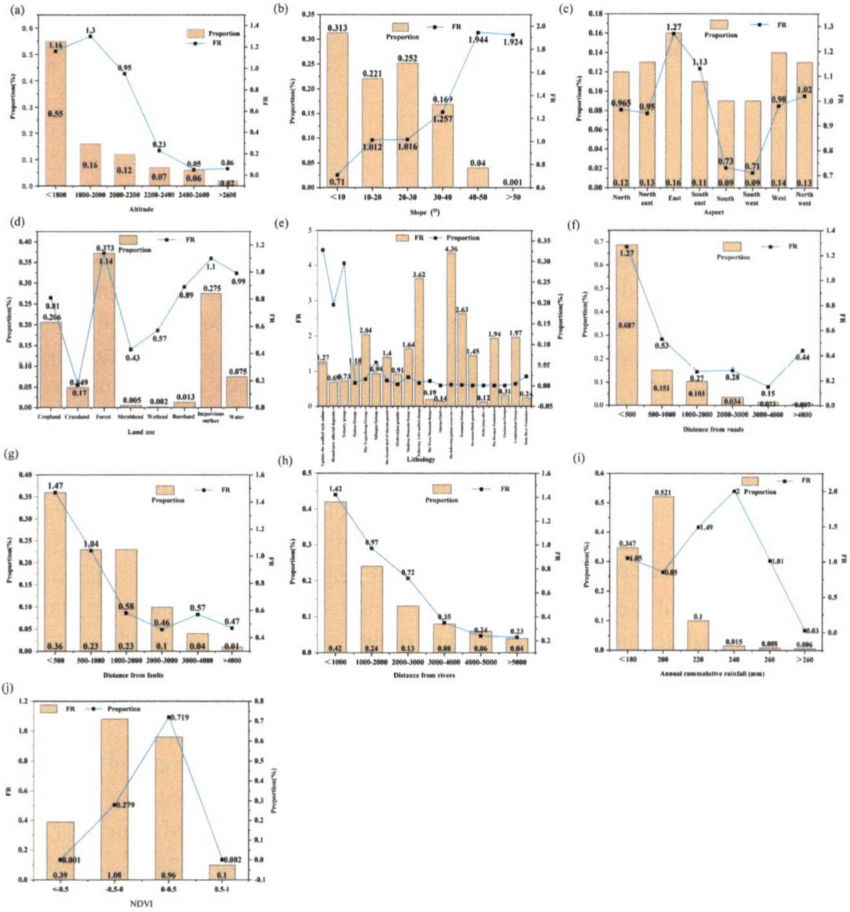

Fig. 5.13 Frequency ratio result

and land is complex. While pursuing economic development, the government should also strengthen the supervision of key human activities such as mining, construction and land reduction to control the degree of damage they cause to the land.

(3) Strengthen infrastructure construction: As an important transportation hub in Northwest China, Lanzhou should intensify the construction and maintenance of infrastructure such as roads and Bridges, especially by regularly reinforcing and renovating existing buildings and facilities to prevent huge damage and indirect economic losses caused by traffic disruptions in the event of disasters.

(4) Strengthen geological disaster monitoring: Actively promote the establishment of a complete geological disaster monitoring and early warning system to promptly identify potential geological hazards. Conduct geological disaster

safety hazard inspections in key districts and counties, and then take corresponding emergency measures to reduce the harm of geological disasters to the public.

5.4 Summary

This chapter introduces an integrated evaluation method for landslide susceptibility based on dynamic and static features combined with InSAR based ground surface deformation. This chapter comprehensively elaborates the experimental technical process from four aspects: data processing, model construction, model training, and accuracy verification. In data processing, it consists of two parts: dynamic data processing and static data processing. The dynamic data used in this paper is the SABA-InSAR surface time series cumulative deformation obtained in the previous text. However, due to atmospheric interference and flight orbit factors, the time intervals of the obtained data are not equal. Therefore, data interpolation is carried out first, and the result is the time series data with equal intervals that satisfy one scene every 12 days. Then, in order not to change the distribution characteristics of the time series and meet the data normalization requirements of the deep learning model, the dynamic normalization formula of the time series is proposed to complete the dynamic data processing; In terms of static data, the 10 landslide influencing factors constructed in the previous text were selected, and the histogram equalization processing was completed to make it still have rich characteristics in a single band. Finally, a total of 24 scenes of temporal cumulative deformation values and 10 landslide influence factors were selected quarterly to jointly construct the overall data set, and the training set, verification set and test set were divided in an 8:2:1 ratio. The joint feature utilization model is constructed by integrating parallel networks. Dynamic features are learned through TD-CNN and Bi-GRU, and static features are obtained through MSCNN to acquire multi-scale features of deep and shallow layers. Finally, the two types of features are fused and learned through channel fusion. From the perspective of model validation, this paper sets up a model that only utilizes static influence factor data as a comparative study. The training curve of the validation set is used to represent the differences in the training process. On the test set, indicators such as the confusion matrix and ROC are used to quantitatively compare the two types of models. The experimental results show that the joint feature utilization model is superior to the model that only learns static features in all aspects. Furthermore, more accurate and reliable evaluation results of landslide susceptibility can be obtained. The trained model was used to evaluate the probability of landslide occurrence pixel by pixel in the study area to produce the evaluation result map of landslide susceptibility. The method proposed in this paper is more accurate in characterizing areas with extremely high susceptibility levels, covering more historical landslide points in a smaller area, and the results are more reliable. Finally, based on the application frequency ratio method in the highly prone areas of the susceptibility mapping results to connect each influencing factor, analyze the contribution level of environmental factors to landslides, and then put forward effective suggestions for the prevention and control of landslides and geological disasters in the main urban area of Lanzhou.

References

Achu, A. L., Aju, C. D., Di Napoli, M., Prakash, P., Gopinath, G., Shaji, E., & Chandra, V. (2023). Machine-learning based landslide susceptibility modelling with emphasis on uncertainty analysis. *Geoscience Frontiers, 14*, Article 101657. https://doi.org/10.1016/j.gsf.2023.101657

Archer, K. J., & Kimes, R. V. (2008). Empirical characterization of random forest variable importance measures. *Computational Statistics & Data Analysis, 52*(4), 2249–2260. https://doi.org/10.1016/j.csda.2007.08.015

Chauhan, S., Sharma, M., Arora, M. K., & Gupta, N. K. (2010). Landslide susceptibility zonation through ratings derived from Artificial Neural Network. *International Journal of Applied Earth Observation and Geoinformation, 12*(5), 340–350. https://doi.org/10.1016/j.jag.2010.04.006

Gao, B., He, Y., Chen, X., Chen, H., Yang, W., & Zhang, L. (2024). A deep neural network framework for landslide susceptibility mapping by considering time-series rainfall. *IEEE Journal of Selected Topics in Applied Earth Observations and Remote Sensing, 17*, 5946–5969. https://doi.org/10.1109/JSTARS.2024.3370218

Gao, B., He, Y., Chen, X., Zheng, X., Zhang, L., Zhang, Q., & Lu, J. (2023). Landslide risk evaluation in Shenzhen based on stacking ensemble learning and InSAR. *IEEE Journal of Selected Topics in Applied Earth Observations and Remote Sensing, 16*, 1–18. https://doi.org/10.1109/JSTARS.2023.3291490

Jiang, Y., Wang, W., Zou, L., & Cao, Y. (2024). Regional landslide susceptibility assessment based on improved semi-supervised clustering and deep learning. *Acta Geotechnica, 19*, 509–529. https://doi.org/10.1007/s11440-023-01950-0

Luna-Alvarez, A., Mujica-Vargas, D., Matuz-Cruz, M., Vianney Kinani, J. M., & Ramos-Diaz, E. (2020). Self-driving through a time-distributed convolutional recurrent neural network. In: *2020 17th international conference on electrical engineering, computing science and automatic control (CCE)* (pp. 1–6). IEEE. https://doi.org/10.1109/CCE50788.2020.9299181

O'Brien, R. M. (2007). A caution regarding rules of thumb for variance inflation factors. *Quality & Quantity, 41*(5), 673–690. https://doi.org/10.1007/s11135-006-9018-6

Pham, B. T., Bui, D. T., Pourghasemi, H. R., Indra, P., & Dholakia, M. B. (2017a). Landslide susceptibility assessment in the Uttarakhand area (India) using GIS: A comparison study of prediction capability of naïve Bayes, multilayer perceptron neural networks, and functional trees methods. *Theoretical and Applied Climatology, 128*, 255–273. https://doi.org/10.1007/s00704-015-1702-9

Pham, B. T., Bui, D. T., Prakash, I., & Dholakia, M. B. (2017b). Hybrid integration of Multilayer Perceptron Neural Networks and machine learning ensembles for landslide susceptibility assessment at Himalayan area (India) using GIS. *CATENA, 149*, 52–63. https://doi.org/10.1016/j.catena.2016.09.007

Sun, C., Wu, Z., Lv, Z., Yao, N., & Wei, J. (2013). Quantifying different types of urban growth and the change dynamic in Guangzhou using multi-temporal remote sensing data. *International Journal of Applied Earth Observation and Geoinformation, 21*, 409–417. https://doi.org/10.1016/j.jag.2011.12.012

Wang, Y., Fang, Z., & Hong, H. (2019). Comparison of convolutional neural networks for landslide susceptibility mapping in Yanshan County, China. *Science of the Total Environment, 666*, 975–993. https://doi.org/10.1016/j.scitotenv.2019.02.263

Zhao, Z., He, Y., Yao, S., Yang, W., Wang, W., Zhang, L., & Sun, Q. (2022). A comparative study of different neural network models for landslide susceptibility mapping. *Advances in Space Research, 70*(2), 383–401. https://doi.org/10.1016/j.asr.2022.04.055

Chapter 6
Multi-model Integrated Landslide Susceptibility Mapping Method Based on Static-Dynamic Spatiotemporal Data Fusion

Traditional landslide susceptibility mapping predominantly relies on static conditioning factors, often neglecting dynamic influences, which leads to reduced accuracy and limited reliability in risk estimation. To address this limitation, we propose a multi-model integrated mapping approach that incorporates both static and dynamic spatiotemporal data. By leveraging the complementary characteristics of static and dynamic landslide features, this method aims to enhance the overall reliability of landslide risk evaluations.

6.1 Study Area and Data

6.1.1 Study Area

The study area is Shenzhen, located in the southeastern part of Guangdong Province, China, on the eastern shore of the Pearl River Estuary. It is bordered by Daya and Dapeng Bay to the east, faces the New Territories of Hong Kong across the Shenzhen River to the south, and is adjacent to Dongguan and Huizhou to the north (Fig. 6.1). Shenzhen is recognized as one of China's core economic hubs and international cities, as well as one of the three national financial centers. The total land area is approximately 1997.47 km^2, centered at 22.45°N, 113.76°E. The city experiences a South Asian tropical marine monsoon climate, with the rainy season occurring from April to September each year, during which landslides, floods, and short-duration severe convective storms are common.

Topographically, Shenzhen is characterized by a southeast-high to northwest-low gradient, comprising diverse landforms such as low hills, terraces, low mountains, stepped terrains, and plains. The region is intersected by 24 geological faults, primarily trending northeast and northwest. Due to its coastal setting, the subsurface geology includes materials such as organic sand, silt, and gravel, resulting in

Y. He, *InSAR and Deep Learning in Landslides Research: Intelligent Identification, Risk Assessment and Susceptibility Mapping*, https://doi.org/10.1007/978-981-96-9132-6_6

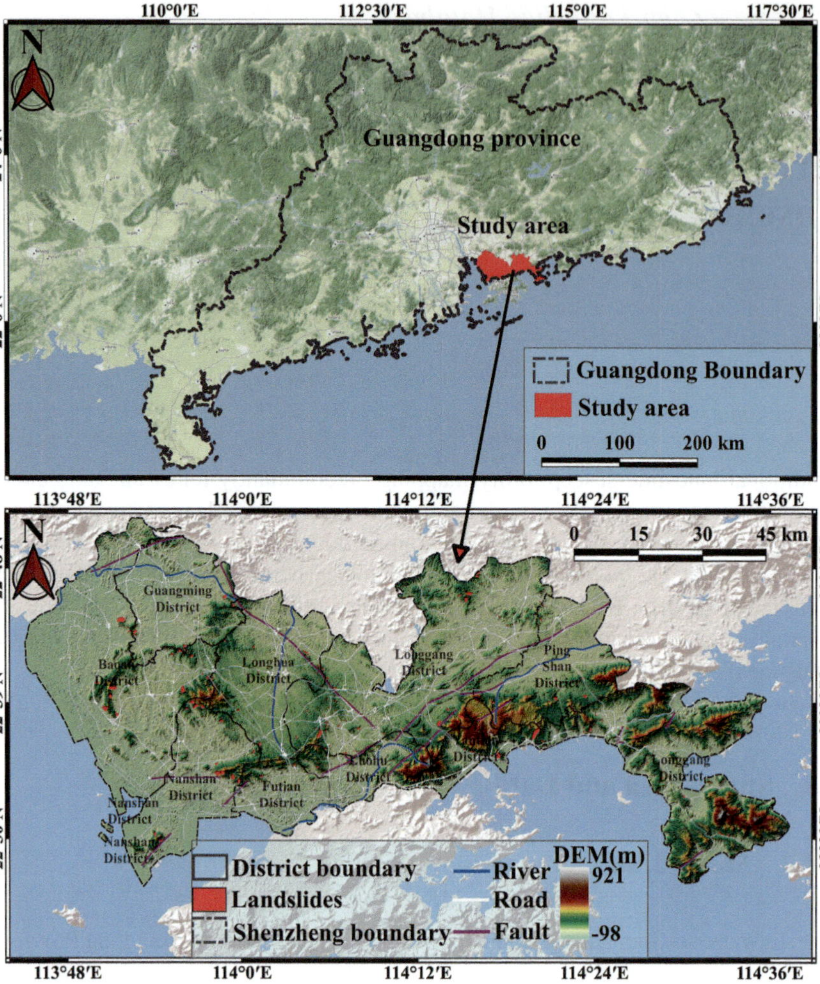

Fig. 6.1 Location and general features of the study area, reprinted from Ref. Gao et al. (2024), copyright 2025, with permission from IEEE

loosely consolidated soils that are highly susceptible to landslides under intense rainfall conditions. According to historical records, a total of 438 landslide events have occurred in Shenzhen (Gao et al., 2024). Notably, on September 17, 2023, continuous heavy rainfall induced by a typhoon triggered multiple landslides in the Wutong Mountain area. Given the severity of this event, comprehensive landslide mapping and corresponding risk management strategies are critically important for Shenzhen.

6.1.2 *Shenzhen Landslide Database*

6.1.2.1 InSAR Deformation Data

InSAR technology has emerged as a powerful tool for monitoring regional ground surface deformation and facilitating the early detection of landslides, providing a novel perspective for accurate and near real-time landslide investigations (Cai, 2024; Liu et al., 2024a, 2024b; Yang et al., 2025). Numerous studies have demonstrated the reliability of InSAR in detecting and monitoring unstable slopes, confirming its efficacy as a dependable technique for landslide mapping (Dai et al., 2022).

SBAS-InSAR, a form of differential interferometric radar technique, utilizes temporally coherent SAR imagery to derive surface deformation through differential processing. This approach is characterized by high sensitivity and broad spatial coverage (Zhang et al., 2020). By selecting SAR image pairs with short temporal and spatial baselines, the SBAS technique effectively reduces the impact of atmospheric disturbances and topographic errors. Through the measurement of phase differences in radar signals acquired at multiple time points, SBAS-InSAR enables the detection of subtle ground deformations with high temporal resolution (Liu et al., 2024a, 2024b).

The ground surface deformation dataset employed in this study was derived from Sentinel-1A SAR imagery, as detailed in Table 6.1. Provided by the ESA since its launch in 2014, Sentinel-1A offers all-weather, day-and-night Earth observation capabilities with cloud-penetrating radar, high spatial resolution, and frequent temporal revisits. These features make it a key resource for scientific investigations, environmental monitoring, and disaster management.

A total of 246 Sentinel-1A scenes covering the Shenzhen from January 2018 to December 2021 across two acquisition tracks were processed using the SBAS-InSAR approach implemented in SARscape. The data processing process consists of the following steps: connection graph generation, interferogram processing, orbital refinement and topographic flattening, and deformation inversion.

(1) Connection Graph Generation: The SBAS methodology begins by constructing a temporal connection graph to link adjacent SAR acquisitions based on coherence estimation. For a set of $N + 1$ images, coherence is calculated for each pair to assess phase correlation. A threshold is then applied to define valid connections, forming a graph where each node corresponds to a SAR image and edges represent coherent image pairs.

Table 6.1 Sentinel-1 data sources and information presentation

Orbital number	Direction	Time	Angle	Heading	Resolution
P:11/F:107	Ascending	2018.01.06–2021.12.28	44.21°	347.63°	5m × 20m
P:113/F:71	Ascending	2018.01.01–2021.12.23	32.93°	347.61°	5m × 20m

(2) Interferogram Processing: This stage includes core operations such as phase computation, filtering, phase removal, and unwrapping. Initially, phase signals are extracted to detect deformation or elevation changes. Filtering and atmospheric correction techniques are employed to suppress noise and eliminate non-deformation signals, particularly those introduced by atmospheric delay and terrain. Phase unwrapping resolves the inherent ambiguity in phase values to enable consistent interpretation.
(3) Orbital Refinement and Flattening: Orbital refinement enhances spatial accuracy by correcting geometric misalignments, while flattening eliminates phase ramps caused by topographic relief. These corrections are supported by external DEM and ground control points (GCPs), which are selected based on criteria such as non-deforming areas with high coherence, smooth phase continuity, and minimal residual phase artifacts.
(4) Deformation Inversion: SBAS-InSAR includes a two-step inversion process. The first inversion estimates initial deformation rates and residual topographic signals using temporal derivatives and singular value decomposition (SVD). Least-squares fitting of the time series provides initial velocity estimates. In the second inversion, high-pass and low-pass filtering are applied in the spatial and temporal domains to remove atmospheric and noise-induced components. The final outputs include line-of-sight (LOS) deformation velocities and cumulative displacement time series. A geocoding step then transforms these results into geographic coordinates consistent with the DEM reference for subsequent analysis.

6.1.2.2 Landslide Inventory Data

With the advancement of InSAR technology, its application in ground deformation monitoring and landslide detection has become increasingly widespread (Dai et al., 2022; He et al., 2023). Leveraging the dynamic observation capabilities of SBAS-InSAR, Dai et al. used SBAS-InSAR technology to monitor subtle displacements on reservoir banks and, in combination with a geological evaluation model, conducted a landslide susceptibility assessment for the Baihetan Dam (Dai et al., 2023). Field validation confirmed that the method offers effective and accurate dynamic landslide susceptibility mapping. In addition, InSAR based ground deformation technology are used to assist in identifying potential landslides, and the method was successfully applied in Lanzhou (Zhao, 2024).

Shenzhen has recorded over 400 landslides historically. However, due to rapid urbanization and land-use modifications, some historical landslides have disappeared. In this study, InSAR-derived ground surface deformation data were used as an indicator of latent landslide. These data were integrated with landslide-prone zones identified in the 2018–2021 Shenzhen Geological Disaster Prevention and Control Plan (https://pnr.sz.gov.cn/, accessed on May 6, 2025), along with information on the spatial distribution, morphology, vegetation coverage, and environmental context of historical landslides. High-resolution historical imagery from Google

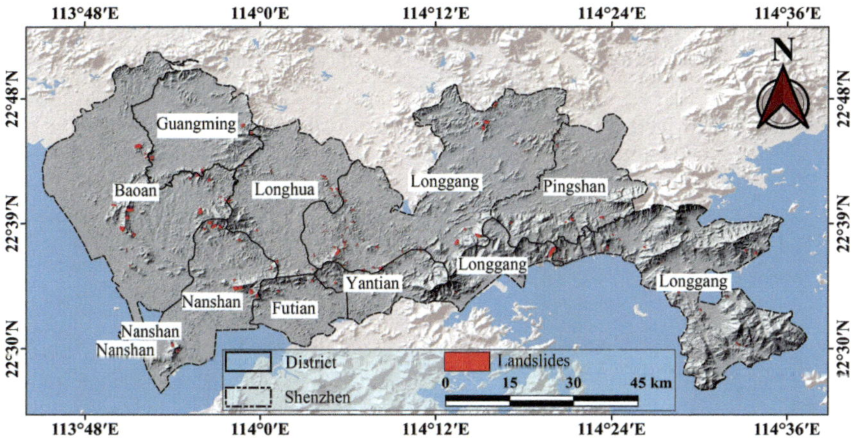

Fig. 6.2 Landslide detection results based on InSAR deformation constraints, reprinted from Ref. Gao et al. (2024), copyright 2025, with permission from IEEE

Earth was employed to delineate landslide samples, resulting in the identification of 205 landslide bodies (Fig. 6.2), comprising both relict and actively deforming potential landslides.

6.2 Landslide Influencing Factors in Shenzhen

6.2.1 Static Landslide Influencing Factors

Static environmental factors play a pivotal role in landslide susceptibility mapping. Given that landslides typically occur in regions with similar geomorphological and environmental characteristics, extracting environmental features from known landslide sites enables the identification of analogous areas at risk. This approach is crucial for scientifically evaluating potential landslide and implementing appropriate mitigation strategies.

Landslide occurrence is primarily controlled by a combination of topographic, geological, hydrological, and anthropogenic factors. Based on the environmental context of Shenzhen, this study selected 12 representative static variables: elevation, slope, aspect, surface roughness, terrain relief, curvature, distance to faults, distance to rivers, land use types, distance to roads, soil type, and sand content (Chen et al., 2017). The data sources and parameters are detailed in Table 6.2.

Elevation was derived from a 30 m resolution DEM, from which several topographic metrics—slope, aspect, curvature, surface roughness, and terrain relief—were subsequently calculated. Among these, slope is considered a critical factor affecting slope stability (Lv et al., 2022), as it influences both infiltration processes

Table 6.2 Sources and detailed parameters of static landslide impact factors

Factors	Data source	Resolution/ scale	Classification criteria	Classification methods
Elevation (m)	DEM (http://earthdata.nasa.gov) accessed on May 6, 2025	30 m	(1) 0 to 38;(2) 38 to 98;(3) 98 to 186;(4) 186 to 302;(5) 302 to 464;(6) 464 to 921	Natural breakpoint method
Slope(°)			(1) 0 to 4.78 (2) 4.78 to 10.16; (3) 10.16 to 16.74; (4) 16.74 to 24.51; (5) 24.51 to 33.78; (6) 33.78 to 76.22	Natural breakpoint method
Aspect			(1) Plane; (2) North; (3) Northeast; (4) East; (5) Southeast; (6) South; (7) Southwest; (8) West, (9) Northwest; (10) North	Equally spaced
Surface roughness			(1) 1 to 1.04 (2) 1.04 to 1.14 (3) 1.14 to 1.33 (4) 1.33 to 4.20	Natural breakpoint method
Surface relief			(1) 0 to 3 (2) 3 to 7 (3) 7 to 13 (4) 13 to 20 (5) 20 to 104	Natural breakpoint method
Curvature			(1) −70 to −4.95 (2) −4.95 to −1.02 (3) −1.02 to 1.78 (4) 1.78 to 5.14 (5) 5.14 to 73	Natural breakpoint method
Distance from fault (m)	http://www.resdc.cn accessed on May 6, 2025	30 m	(1) 0 to 3177 (2) 3177 to 6354 (3) 6354 to 9531 (4) 9531 to 12,708 (4) 12,708 to 15,885	Equally spaced
Distance from river (m)	https://www.webmap.cn accessed on May 6, 2025		(1) 0 to 6367 (2) 6367 to 12,734 (3) 12,734 to 19,101 (4) 19,101 to 25,468 (5) 25,468 to 31,835	Equally spaced
Distance from road (m)	–	–	(1) 0 to 6972 (2) 6972 to 13,944 (3) 13,944 to 20,917 (4) 20,917 to 27,899 (5) 27,899 to 34,862	Equally spaced

(continued)

Table 6.2 (continued)

Factors	Data source	Resolution/ scale	Classification criteria	Classification methods
LUT	https://viewer.esa-worldcover.org/worldcover/ accessed on May 6, 2025	10 m	(1) Woodland; (2) Shrubs; (3) Grassland (4) Cropland; (5) Buildings; (6) Bare Ground/Sparse Woodland; (7) Permanent Water Bodies; (8) Wetlands	Oversight classification
Soil types	Nanjing Soil Research Institute, Chinese Academy of Sciences	1:4 million	(1) low-activity Fe-strong acidic soil; (2) simple-breeding saline soil; (3) simple-breeding low-activity strongly acidic soil; (4) thin-layer soil	–
Sand content	Data Center for Resource and Environmental Sciences, Chinese Academy of Sciences	1:1 million	(1) 25 to 30 (2) 30 to 34 (3) 34 to 45 (4) 45 to 60 (5) 60 to 76	Natural breakpoint method

and the spatial distribution of shear stress (Chen et al., 2017). The aspect and the length-to-width ratio may also indirectly influence slope instability by modulating solar radiation exposure (Tien Bui et al., 2012). Surface roughness, calculated as the ratio between the actual terrain surface area and its horizontal projection (Eq. (6.1)), quantifies micro-topographic variability and erosion susceptibility—higher values typically indicate more fragmented terrain and severe surface degradation. Terrain Ruggedness Index (TRI) further reflects topographic complexity, derived from local elevation differences within a specified area (Gao et al., 2023).

$$TRI = \frac{1}{\cos(tan\beta \times \frac{\pi}{180})} \tag{6.1}$$

Within the study area, fault activity is characterized by regional faults in Guangdong Province and compressional fault zones comprising two main fault belts, with strike-slip angles between 30° and 50°, indicative of medium- to high-angle structural features. Proximity to these faults often correlates with discontinuous lithology and loose soil conditions, increasing the likelihood of slope failure (Zhang et al., 2022). River networks exert influence through hydraulic erosion and sediment removal, contributing to terrain instability (Chen et al., 2021). Human-induced factors, such as land use patterns and proximity to roads, serve as indicators of anthropogenic disturbance (Wang & Xu, 2017). Soil-related variables, including soil type and sand content, are essential for assessing subsurface hydromechanical behavior. Differences in density, porosity, shear strength, and permeability affect the soil's capacity to resist failure. In particular, elevated pore water pressure reduces effective stress, thereby decreasing shear strength and increasing landslide risk (Sameen et al., 2020). All 12 static conditioning factors used in this study are illustrated in Fig. 6.3.

6.2.2 *Dynamic Landslide Influencing Factors*

(1) Time-series rainfall as a dynamic landslide trigger
Rainfall is a primary factor contributing to slope instability (He et al., 2010). Previous studies have emphasized that both rainfall intensity and temporal patterns play a crucial role in determining landslide dynamics and slope failure mechanisms (Hong et al., 2019). Variations in rainfall intensity, duration, and temporal evolution can differentially influence slope stability (He et al., 2023). As illustrated in Figs. 6.4a and b, rainfall typically elevates pore water pressure, thereby reducing the effective stress and shear strength of slope materials, ultimately resulting in failure (Xing et al., 2021). In addition, rainfall-induced infiltration alters the hydrological regime of slopes and promotes physical and chemical weathering processes—such as softening, swelling, and disintegration—that reduce soil strength. Surface processes such as runoff, erosion, and dissolution also contribute to slope morphology changes and mass redistribution, increasing gravitational loading and driving landslide occurrence (Wu et al., 2022). In Shenzhen, where rainfall is a dominant landslide-inducing factor (Zhou et al., 2021), intense short-duration storms significantly elevate soil moisture, reduce slope stability, and amplify landslide risk (Yin et al., 2016). Therefore, analyzing the spatiotemporal evolution of rainfall through time series is vital to understanding and predicting landslide dynamics.
This study utilized 48 monthly rainfall observations from 2018 to 2021 as dynamic temporal indicators. The data, sourced from the Loess Plateau Science Data Center under the National Earth System Science Data Sharing Platform of China, have a spatial resolution of 1 km and are updated annually (Luo et al., 2019). To ensure spatial consistency, rainfall datasets were resampled to 30 m resolution using inverse distance weighting (IDW) interpolation in QGIS. Figure 6.5a–i present the spatiotemporal distribution of rainfall during the study period, while Fig. 6.5m illustrates the temporal range of minimum and maximum rainfall. A clear seasonal fluctuation pattern was observed, with peak rainfall typically occurring between March and October, highlighting the extended rainy season in the region.

(2) Time-series NDVI for assessing vegetation dynamics in landslide areas
Shenzhen is characterized by abundant vegetation coverage, which plays a critical role in mitigating soil erosion and maintaining slope stability. Vegetation dynamics are closely linked to landslide processes. Prior to landslide occurrence,

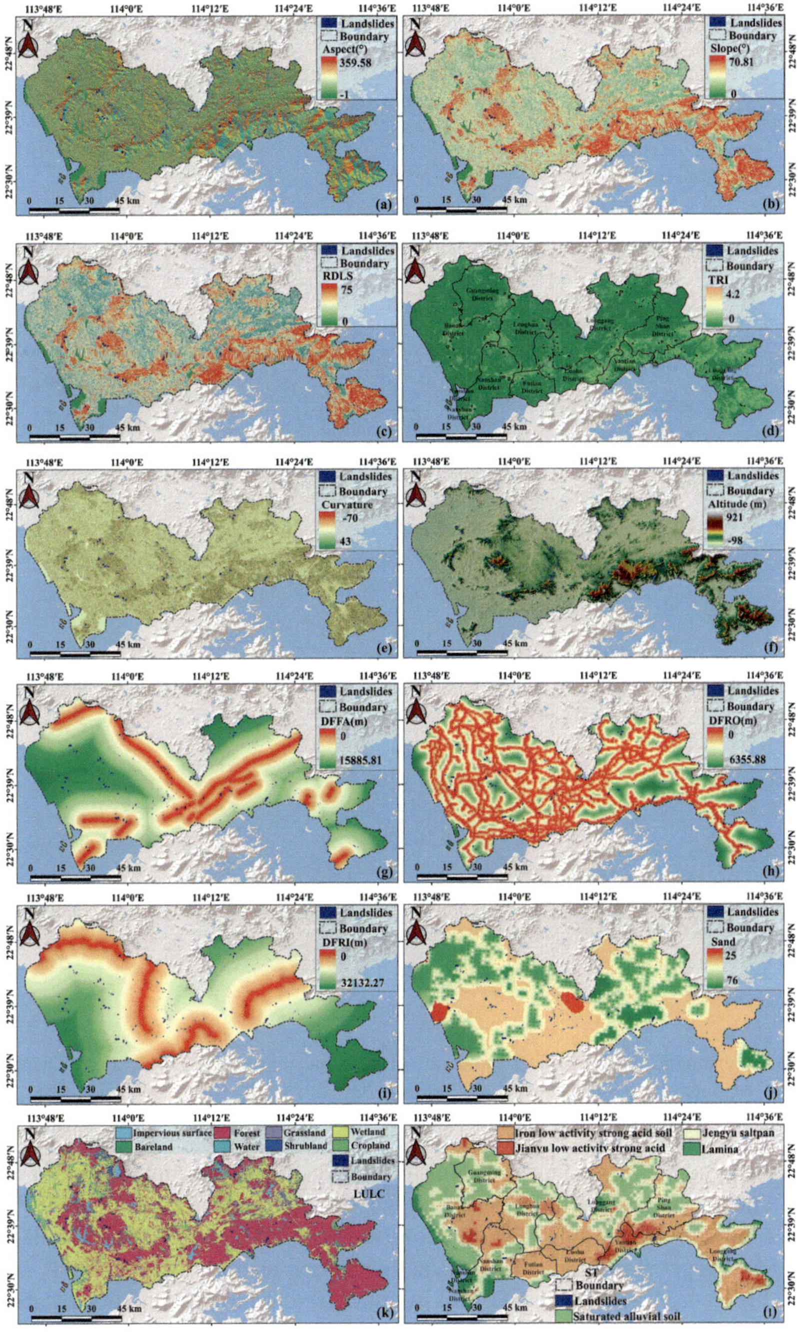

Fig. 6.3 Static landslide influencing factors: **a** aspect, **b** slope, **c** surface roughness, **d** surface relief, **e** curvature, **f** elevation, **g** distance to fault, **h** distance to road, **i** distance to river, **j** sand content, **k** land use, **l** soil types, reprinted from Ref. Gao et al. (2024), copyright 2025, with permission from IEEE

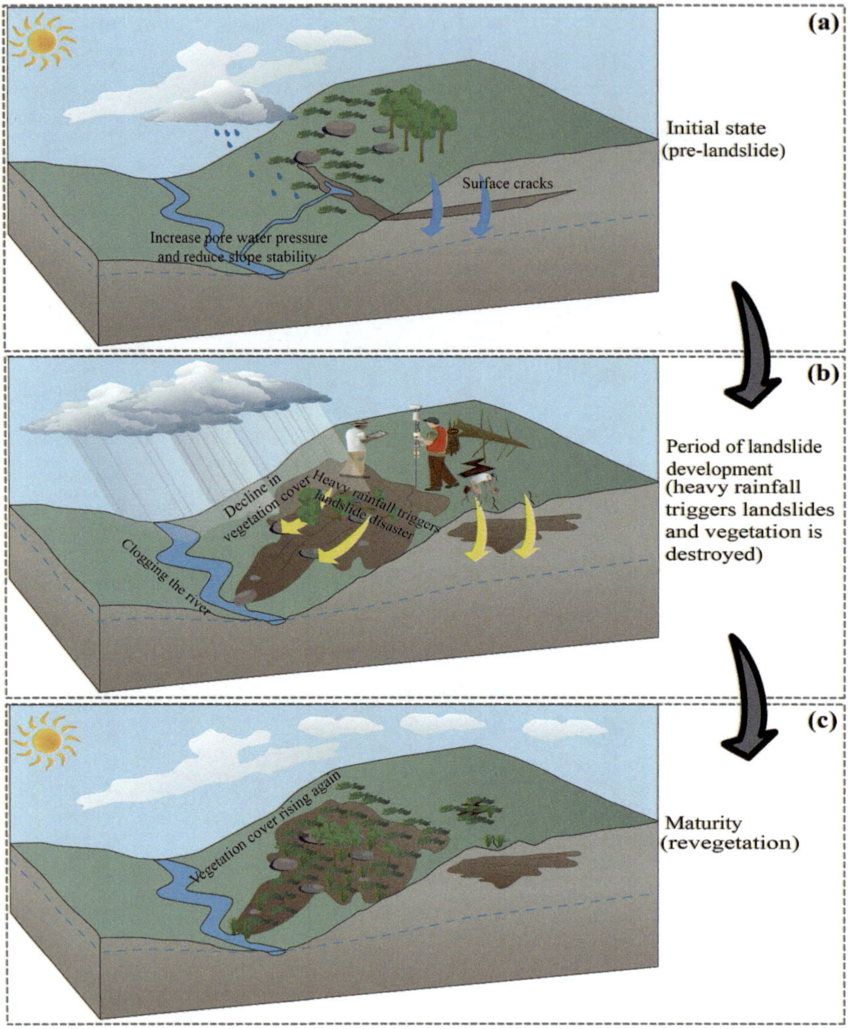

Fig. 6.4 Schematic diagram of the influence of dynamic influencing factors on the landslide process, reprinted from Ref. Gao et al. (2024), copyright 2025, with permission from IEEE

soil structure on the landslide body is often disrupted, leading to reduced vegetation cover and a declining NDVI trend in the affected area (Zhao et al., 2022). A sudden drop in NDVI is often observed during landslide events, reflecting rapid vegetation loss and significant landscape alteration (Hong et al., 2016) (Fig. 6.4a, b). As landslide activity diminishes and soil conditions stabilize, vegetation tends to recover, as supported by findings from Xiang et al. (2023), with NDVI values gradually increasing again over time (Fig. 6.6m). Given Shenzhen's favorable climate and rainfall, post-landslide vegetation regrowth is

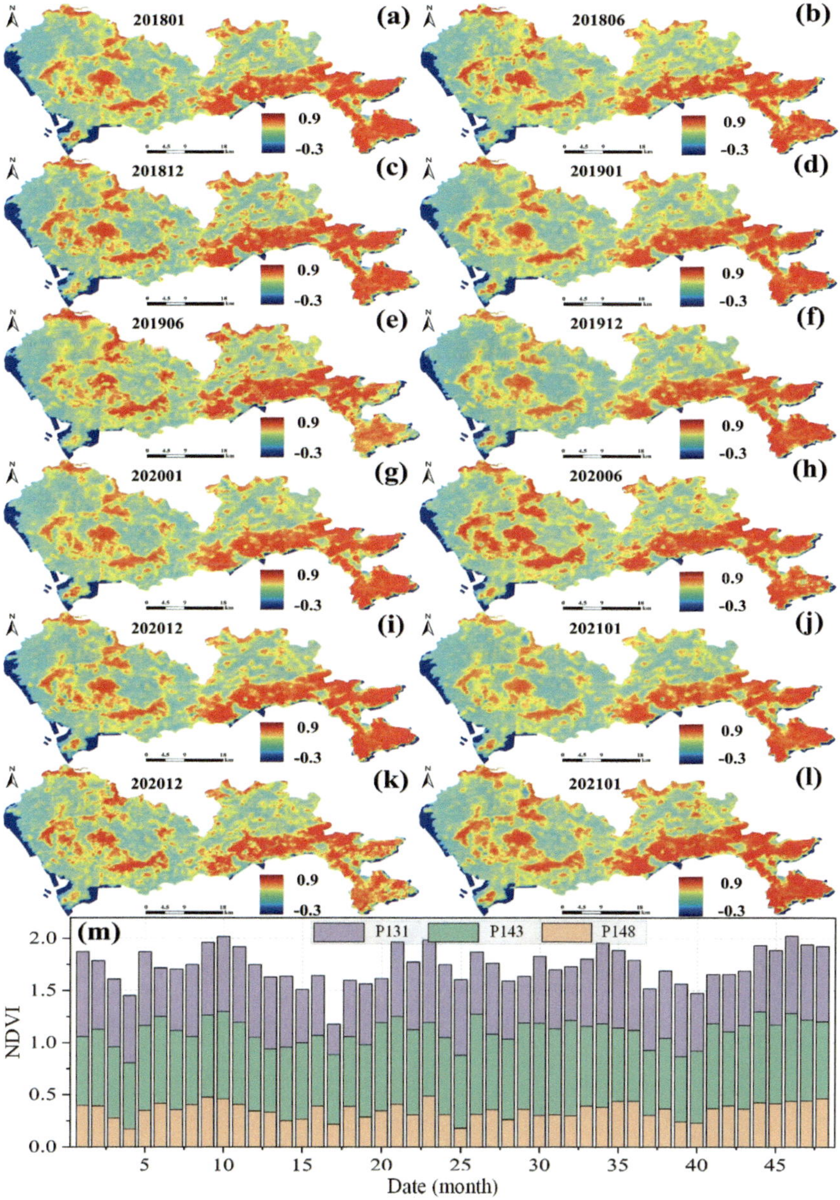

Fig. 6.5 Time series precipitation in Shenzhen from 2018 to 2021, reprinted from Ref. Gao et al. (2024), copyright 2025, with permission from IEEE

typically evident, reinforcing the utility of NDVI time series in tracking landslide evolution.

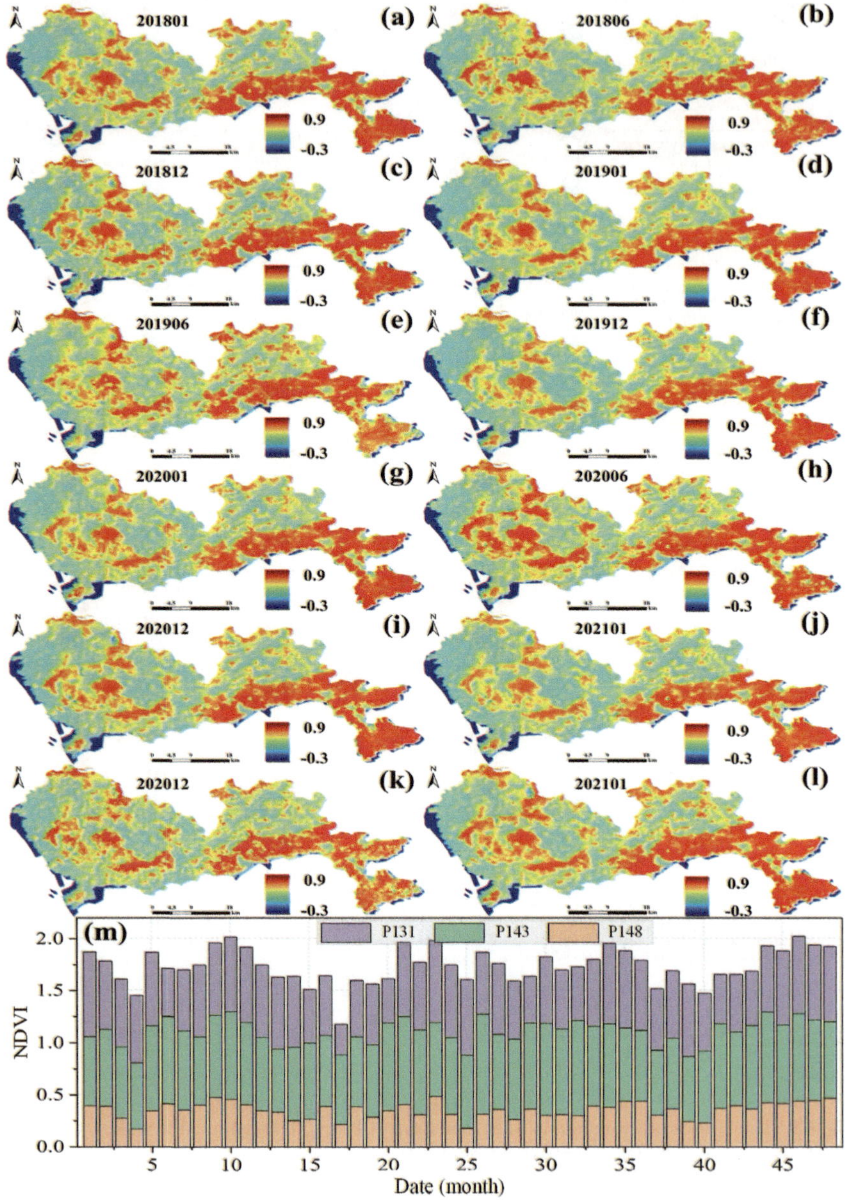

Fig. 6.6 Time series NDVI of Shenzhen in 2018–2021, reprinted from Ref. Gao et al. (2024), with permission from IEEE

In this study, monthly MODIS 13Q NDVI products from 2018 to 2021 were utilized, with a native spatial resolution of 250 m (https://modis.gsfc.nasa.gov, accessed on May 6, 2025). These were resampled to 30 m using inverse distance weighting to facilitate integration with other spatial datasets. To evaluate vegetation dynamics in relation to landslides, three historically landslide-prone zones—P131, P143, and P148—were selected. Figure 6.6a–l display the spatiotemporal NDVI variations across the study period, while Fig. 6.6m illustrates the time series NDVI trajectories for the selected zones. All three areas exhibited a characteristic pattern of NDVI decrease followed by gradual recovery, indicating the potential of vegetation indices to serve as proxies for monitoring landslide recovery and regrowth processes.

6.3 Landslide Susceptibility Mapping Based on Multi-model Fusion of Static–Dynamic Multi-source Data

Based on the analysis of the spatial distribution and developmental characteristics of landslides in Shenzhen, twelve static environmental factors were identified as significantly correlated with landslide occurrences. Building upon this, two time-series dynamic factors were further incorporated, selected according to local landslide typologies, as additional input features. To effectively integrate both static and dynamic indicators, two deep learning-based landslide susceptibility mapping frameworks were proposed, leveraging ensemble learning strategies to improve the reliability of susceptibility prediction.

6.3.1 Research Methodology

6.3.1.1 Collinearity Analysis Method for Landslide Influencing Factors

(1) Multicollinearity analysis

Multicollinearity analysis is conducted to detect strong linear dependencies among independent variables, which can compromise model stability and reliability (Ciampalini et al., 2016). In regression modeling, multicollinearity occurs when two or more predictor variables exhibit significant intercorrelation (Chen et al., 2017), potentially leading to redundancy and interpretational ambiguity. Given that landslide occurrences are influenced by multiple environmental factors, selecting highly correlated variables may result in overlapping explanatory effects. To mitigate this, the Variance Inflation Factor (VIF) and Tolerance Level (TOL) (Zhang et al., 2022) were employed to quantify the degree of multicollinearity. The formulas used are as follows:

$$VIF = \frac{1}{1 - R^2} \tag{6.2}$$

$$TOL = 1 - R^2 \tag{6.3}$$

Here, R^2 denotes the coefficient of determination from the regression of a given explanatory variable on all other predictors. A *VIF* value between 1 and 10 or a *TOL* above 0.1 suggests that multicollinearity is not problematic, whereas *VIF* values exceeding 10 or *TOL* values below 0.1 indicate severe multicollinearity.

The Pearson correlation coefficient is a widely used statistical measure that quantifies the strength and direction of the linear relationship between two continuous variables. It can also provide insight into potential nonlinear dependencies among factors (Ciampalini et al., 2016). The calculation is given in Eq. (6.4):

$$r = \frac{\sum (X_i - \overline{X})(Y_i - \overline{Y})}{\sqrt{\sum (X_i - \overline{X})^2 \sum (Y_i - \overline{Y})^2}} \tag{6.4}$$

In this equation, X_i and Y_i denote the observed values of the independent and dependent variables, while $\overline{X}$ and $\overline{Y}$ represent their respective means. The pearson correlation coefficient r ranges from −1 to 1, indicating a spectrum from perfect negative to perfect positive linear correlation, with 0 signifying no linear relationship.

(2) Feature interpretability analysis method

Although extensive research has been devoted to landslide susceptibility mapping, existing models inevitably exhibit limitations. Data-driven approaches, including both statistical methods and machine learning techniques, offer the capability to process a large number of conditioning factors for landslide susceptibility mapping. Models such as fuzzy logic, logistic regression, random forest, and convolutional neural networks have shown considerable promise by leveraging diverse landslide-related variables as inputs to generate susceptibility maps.

Despite their success, these models often rely heavily on expert-driven selection and classification of input features, which can limit their generalizability and interpretability. Interpretable feature analysis is therefore essential in landslide studies. Traditional approaches, constrained by limited computational power, often fail to disentangle the individual contributions of each conditioning factor to the landslide susceptibility outcomes, nor do they adequately capture the interdependencies among variables.

The ability to interpret the influence of individual landslide conditioning factors allows researchers to more precisely identify hazardous areas. Understanding the effects of terrain, climate, soil, and other environmental variables enhances the ability to implement targeted prevention and mitigation strategies. Furthermore, such interpretability empowers policymakers and stakeholders to better comprehend the rationale behind susceptibility predictions, thereby facilitating more informed decisions regarding land-use planning, early warning systems, and disaster risk reduction measures.

In this study, Partial Dependence Plots (PDPs) are employed to explore the interpretability of selected static landslide conditioning factors and their individual contributions to the model's susceptibility predictions. PDPs illustrate how the average model prediction responds to variations in a single input feature, while holding all other features constant. Specifically, the target feature is sampled across its value range, other variables are fixed, and the mean predicted outcome is calculated for each sample. By visualizing the relationship between the feature and the output, PDPs provide a deeper understanding of both feature influence and interaction patterns. The mathematical formulation of PDP is as follows:

$$PDP_j(x_j) = E_{\mathbf{X}_{\sim j}}\left[f\left(x_j, \mathbf{X}_{\sim j}\right)\right] \tag{6.5}$$

Here, $PDP_j(x_j)$ represents the partial dependence value of the feature x_j; $E_{\mathbf{X}_{\sim j}}$ denotes the expectation taken over all other input features excluding x_j; and $f\left(x_j, X_{\sim j}\right)$ refers to the model's predicted output given the combination of x_j and the remaining features.

6.3.1.2 Landslide Susceptibility Mapping Method Based on Stacking Ensemble Framework

Although CNN, MLP, and GRU models have shown promising performance in landslide susceptibility mapping, they tend to underperform when processing integrated datasets that include both static and large-scale temporal dynamic variables. Such scenarios often lead to model overfitting and redundant information. The ensemble learning paradigm offers significant advantages in model integration and performance enhancement. Among them, stacking—a model-agnostic ensemble strategy—combines the strengths of multiple base learners to improve both predictive accuracy and model robustness (Divina et al., 2018). The key idea is to use the predictions of individual base classifiers as input features for a higher-level meta-learner, which then produces the final prediction (Cui et al., 2021).

In this study, a stacking-based ensemble learning framework was developed to produce landslide susceptibility maps. The architecture consists of two levels: CNN, MLP, and GRU serve as the base learners in the first layer, while a Support Vector Regression (SVR) model functions as the meta-learner in the second layer. The overall workflow is illustrated in Fig. 6.7. Initially, the dataset is partitioned into training and testing subsets. The training data is further divided into five folds, with four folds used to train the base learners and the remaining fold reserved for validation. Through five-fold cross-validation, each base learner generates prediction outputs on the validation sets, which are then treated as new landslide-related features for the second layer.

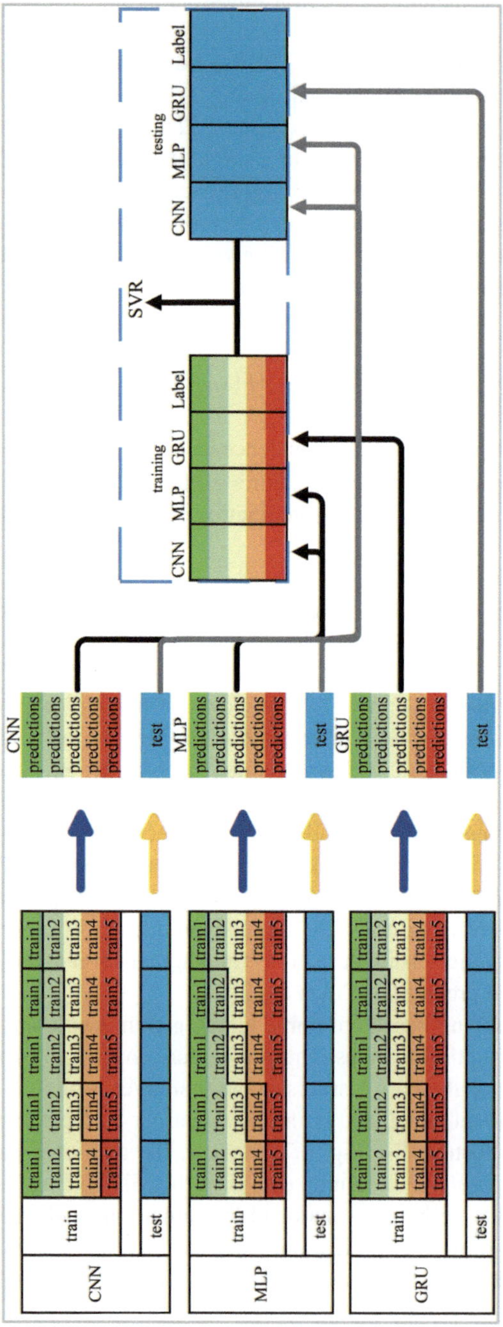

Fig. 6.7 Structure of the stacking ensemble learning model, reprinted from Ref. Gao et al. (2023), copyright 2025, with permission from IEEE

During the meta-learning phase, the outputs from the CNN base learner are concatenated into one-dimensional feature vectors, followed by the corresponding outputs from the MLP and GRU models. This results in an N × 3 feature matrix (A1, A2, A3), which serves as the new training input for the SVR meta-model. Similarly, predictions from the base learners on the test data are combined to form a new N × 3 test matrix (B1, B2, B3). The SVR model is trained on the newly constructed training features and evaluated on the test set to assess the final predictive performance.

6.3.1.3 Landslide Susceptibility Mapping Method Based on TSDNN Ensemble Framework

Conventional neural network architectures often struggle to concurrently process static and temporal dynamic data, limiting their ability to effectively capture the full spectrum of landslide-related characteristics. To overcome this challenge, this study employs a set of high-performance models—including MSCNN, TDCNN, and Bi-LSTM—to construct a deep neural network architecture capable of integrating both static and time-series dynamic landslide predictors. This approach is designed to enhance the reliability and robustness of landslide susceptibility mapping. The proposed Temporal-Static Deep Neural Network (TSDNN) framework is composed of three primary modules (Fig. 6.8): (1) a time-series rainfall factor processing module; (2) a static factor processing module; and (3) a feature fusion and output module.

Part 1: Traditional Bi-LSTM networks are limited in capturing spatial neighborhood dependencies when dealing with datasets that span both spatial and temporal dimensions. As a result, they struggle to construct precise spatial representations or extract local spatial features from time-series landslide inputs. To address this, we propose a three-layer Bi-LSTM framework to effectively extract temporal features from rainfall time series. Prior to being fed into the Bi-LSTM, the rainfall time-series dataset—augmented with temporal and channel dimensions—is first processed by a Temporal Dilated Convolutional Neural Network (TDCNN). This network utilizes multiple convolutional layers with 3 × 3 kernels and increasing depths (32, 64, 128, and 256) to progressively extract dynamic spatial features. The resulting high-level, abstract spatiotemporal feature maps are then transformed via a dense layer into spatiotemporal feature vectors of shape (None, 48, 256), preserving temporal characteristics. These vectors are subsequently input into the multi-layer Bi-LSTM for comprehensive spatiotemporal feature extraction. At the Bi-LSTM output stage, the temporal dimension is eliminated while retaining spatial information, resulting in feature vectors of shape (None, 64).

Part 2: To capture multi-scale features from static spatial data, a Multi-Scale Convolutional Neural Network (MSCNN) is employed. This module is designed to extract both low-level and high-level spatial features through three convolutional layers with kernel sizes of 3 × 3 and depths of 64, 128, and 256, respectively, thereby enabling effective multi-scale feature representation.

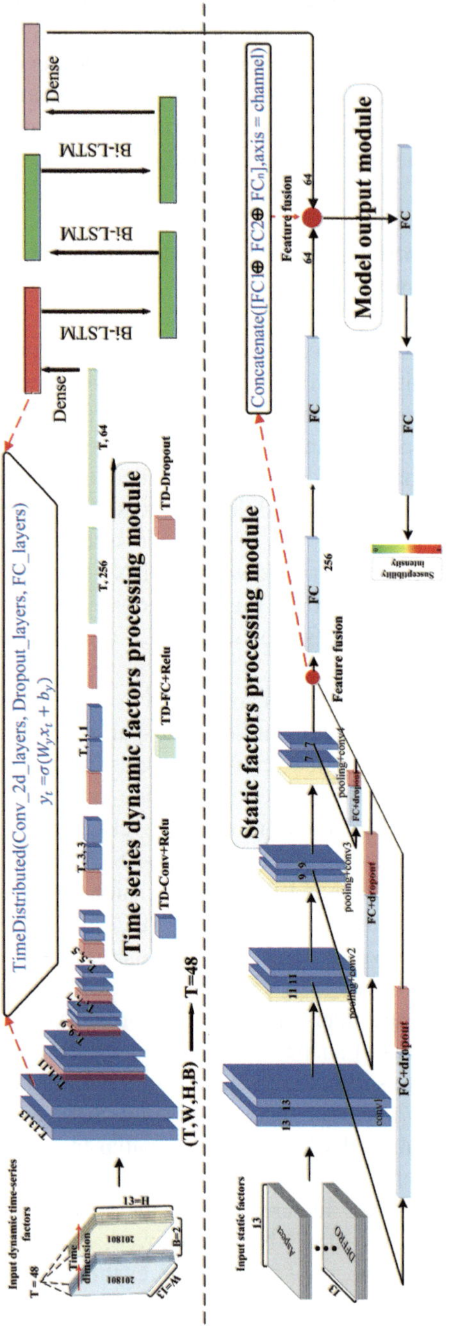

Fig. 6.8 Structure of the TSDNN model, reprinted from Ref. Gao et al. (2024), copyright 2025, with permission from IEEE

Part 3: This module integrates temporal dynamic features and spatial features for final prediction. Assuming the extracted spatial features f_{pk} and temporal features f_{tk} have equal channel dimensions—denoted by X_1, X_2, …, X_c and Z_1, Z_2, …, Z_c, respectively—the fused feature F_f is computed as illustrated in the yellow dashed box of Fig. 6.8. Both the spatial features from MSCNN and the temporal features from TDCNN-Bi-LSTM are projected into 64-dimensional vectors. These vectors are then concatenated to form a 128-dimensional joint feature vector F_f. Finally, two fully connected layers—activated by ReLU and Sigmoid functions, respectively—are applied to produce a landslide susceptibility probability score that captures both spatial and temporal dependencies for downstream analysis.

$$F_f = \sum_{i=1}^{c} X_i * K_i + \sum_{i=1}^{c} Y_i * K_{i+c} \tag{6.6}$$

6.3.2 Collinearity Analysis of Landslide Influencing Factors

6.3.2.1 Collinearity Analysis

The results of the multicollinearity analysis for the 12 selected static landslide conditioning factors are presented in Table 6.3. Among these, the slope and surface roughness exhibit relatively high variance inflation factors (VIFs) of 6.66 and 5.01, respectively, and correspondingly lower tolerance (TOL) values of 0.15 and 0.20. Nevertheless, all variables meet acceptable thresholds for multicollinearity (i.e., VIF $<$ 10 or TOL $>$ 0.1), and significance testing confirms that the predictors fall within the desired confidence interval, indicating that multicollinearity does not pose a substantial concern in this analysis.

To ensure the independence of the 12 selected landslide conditioning factors and to avoid strong correlations or multicollinearity, Pearson correlation analysis was conducted to further examine the relationships between the factors. The results of the correlation analysis, presented in Fig. 6.9, clearly show that the absolute values of the correlation coefficients between the selected factors are all below 0.5, indicating the absence of significant correlations. As a result, the static landslide conditioning factors in this study are considered independent, with no evidence of multicollinearity.

6.3.2.2 Feature Interpretability Analysis

In this study, the PDP method is employed to explain the contributions of selected static landslide conditioning factors and the dependencies between individual factor features and others. As shown in Fig. 6.10, the PDP interpretability analysis curve

Table 6.3 Results of multiple covariance analysis of 3 static impact factors

Static landslide impact factor	Significance	Multicollinearity analysis	
		TOL	VIF
Soil type	0	0.93	1.07
Slope	0	0.15	6.66
Sand content	0	0.97	1.029
Distance to road	0.026	0.93	1.0
Distance to river	0.002	0.90	1.11
Surface relief	0	0.46	2.14
LUT	0.001	0.77	1.28
Distance to fault	0	0.92	1.08
Elevation	0	0.66	1.51
Curvature	0.012	0.99	1.00
Surface roughness	0	0.20	5.01
Aspect	0.252	0.99	1.00

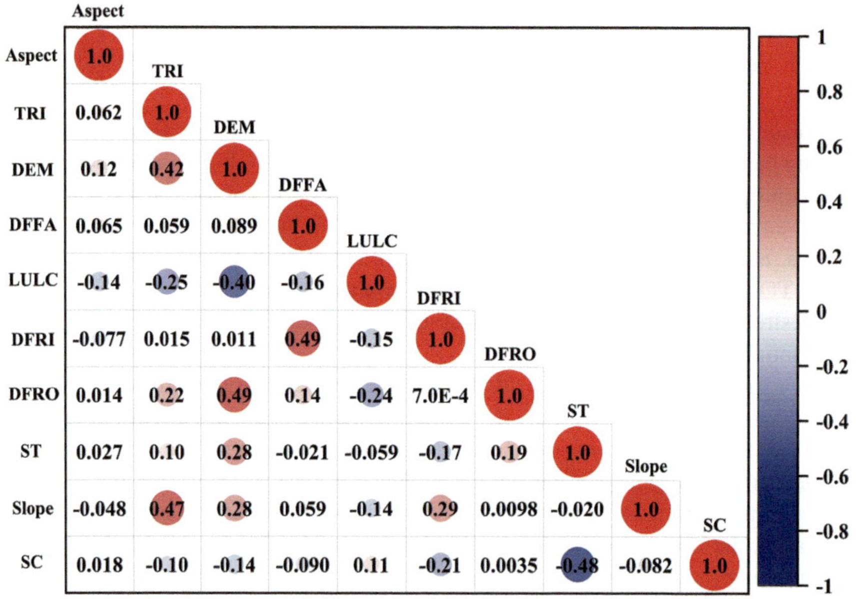

Fig. 6.9 Pearson correlation coefficient analysis, reprinted from Ref. Gao, 2023, copyright 2025, with permission from IEEE

illustrates how changes in feature values influence the average model output. An increasing or decreasing trend in the PDP curve indicates the impact of the feature on the model output. For instance, if the PDP curve shows an upward trend, it suggests that as the feature value increases, the average model output also rises, and vice versa. The slope of the PDP curve can further be used to assess the importance of each feature. A steeper slope indicates a greater impact on the model output, whereas a slope of zero suggests no influence. Specifically, for elevation, the PDP value increases when the feature value is less than 0.3, indicating a positive impact on the model output. However, beyond 0.3, as the feature value increases, the average model output decreases, signifying a negative effect. Slope and curvature exhibit a similar pattern, initially increasing and then decreasing. Aspect and distance to fault also show an increasing trend, suggesting a positive influence on the model output. The PDP value for sand content first decreases and then increases, indicating that low sand content in soil has a minimal effect, while higher sand content enhances the output's influence. Similarly, the effects of soil types and land use types features on the PDP are comparable, revealing that only specific soil and land use types have significant impacts on landslides, while others have lesser effects. Finally, the influence of river and road factors on the PDP value shows similarity, with shorter distances to rivers and roads leading to lower feature values, indicating that proximity to rivers and roads increases the likelihood of landslide occurrences.

6.4 Landslide Susceptibility Mapping in Shenzhen Based on the Stacking Ensemble Framework

6.4.1 Model Performance Analysis

While CNN, MLP, and GRU models demonstrate strong performance in landslide susceptibility mapping, challenges arise due to the large volume of remote sensing data and the complexity of the scenes. Single models, in particular, exhibit certain limitations, such as overfitting and information redundancy. To address these issues, this study employs Stacking ensemble learning for landslide susceptibility mapping and compares its performance with that of individual models (CNN, MLP, GRU, and SVR).

To ensure consistency in input parameters across models, this study simultaneously inputs the 12 selected static landslide influencing factors and 2 dynamic temporal landslide influencing factors into the models. The input dimension size is (13, 13, 108), where 13 represents the patch row and column size, and 108 denotes the number of channels, corresponding to the number of landslide factors. During the map generation process based on the proposed Stacking ensemble learning model, a standard sampling strategy (Fang et al., 2021) is employed to create training and testing datasets. Specifically, 70% of the landslide samples (21,281) are randomly

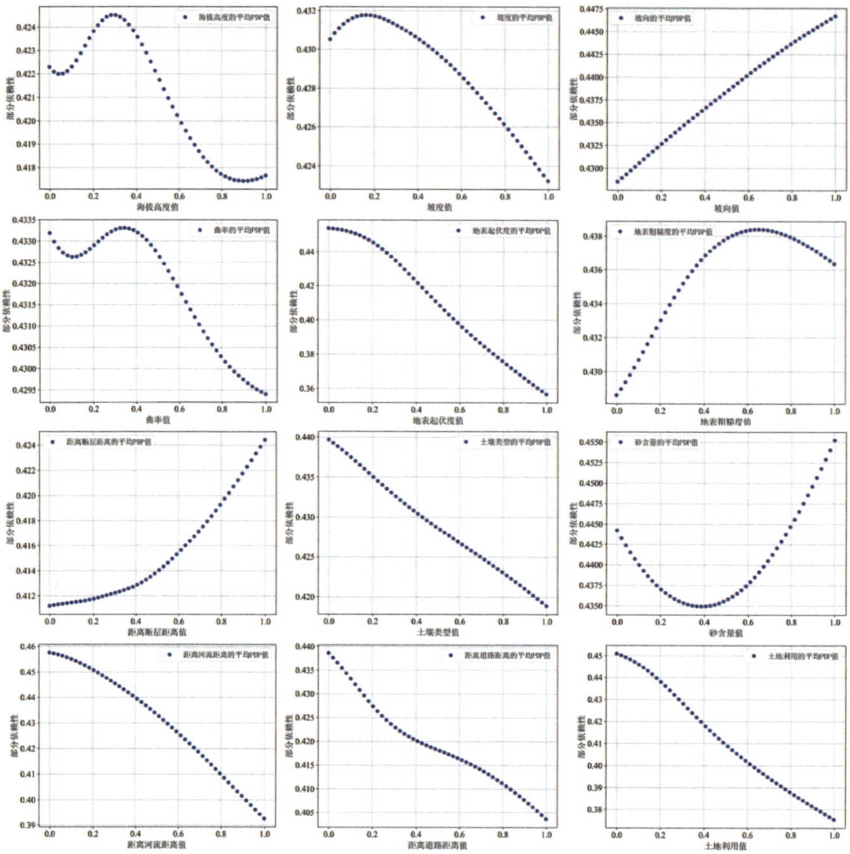

Fig. 6.10 Feature interpretability analysis, reprinted from Ref. Gao et al. (2023), copyright 2025, with permission from IEEE

selected for training, while the remaining 30% (9,120) are used for testing. Additionally, an equal number of non-landslide samples are randomly selected for both the training and testing sets. When training the proposed Stacking ensemble learning model, the study first trains the individual base models and adjusts the hyperparameters using grid search. The hyperparameters of the individual models are then set to match those of the Stacking ensemble model. Furthermore, five-fold cross-validation is applied to determine the optimal hyperparameters for the Stacking model (Luo et al., 2019), as detailed in Table 6.4.

Figure 6.11 presents the training curves for the Stacking ensemble learning model, alongside the comparison models CNN, MLP, and GRU, based on the Logcosh loss function and AUC accuracy values. Specifically, Fig. 6.11(a–f) show the training curves for the CNN, MLP, and GRU models on the same dataset, while Fig. 6.11(g, h) display the loss function and AUC accuracy curves for the proposed Stacking ensemble model after five-fold cross-validation. In the CNN model (Fig. 11a, b), the

Table 6.4 Detailed parameters for model training

Model name	Parameter name	Parameter value
Stacking integration learning model	K-fold	5
CNN model	Batch size	64
	Epochs	120
	Learning rate	0.0004
	Dropout	0.5
	Kernal size	3
	Factor	0.75
MLP model	Batch size	128
	Epochs	120
	Learning rate	0.000127
	Dropout	0.2
	Kernal size	3
	Factor	0.7
GRU model	Batch size	128
	Epochs	120
	Learning rate	0.0000162
	Dropout	0.5
	Kernal size	3
	Factor	0.5
SVR model	Kernel	rbf
	C	1

loss function and AUC accuracy stabilize after 80 iterations, with a Logcosh loss of 0.011 and an AUC accuracy of 0.982 on the validation set. For the MLP model (Fig. 11c, d), to control for variables, the same number of iterations as the CNN model is used. However, the MLP model exhibits a slower convergence rate, with a Logcosh loss of 0.035 and an AUC of 0.969, lower than that of the CNN model. In the GRU model (Fig. 11e, f), convergence is faster, stabilizing after approximately 70 iterations. After 120 iterations, the GRU model achieves a Logcosh loss of 0.034 and an AUC accuracy of 0.969, both lower than those of the CNN model. In contrast, the proposed Stacking ensemble learning model (Fig. 11g, h) integrates the strengths of the base models. As the model iterates, the loss function decreases gradually while the AUC accuracy increases rapidly. After five-fold cross-validation, the Stacking model achieves the lowest Logcosh loss of 0.001 and the highest AUC accuracy of 0.989.

To further evaluate the reliability of the Stacking ensemble learning model, several quantitative metrics based on the confusion matrix, including Recall, Specificity, OA, Kappa, F1 score, and MCC, are used to assess the accuracy of the Stacking model and the individual models. Recall, in particular, measures the proportion of actual

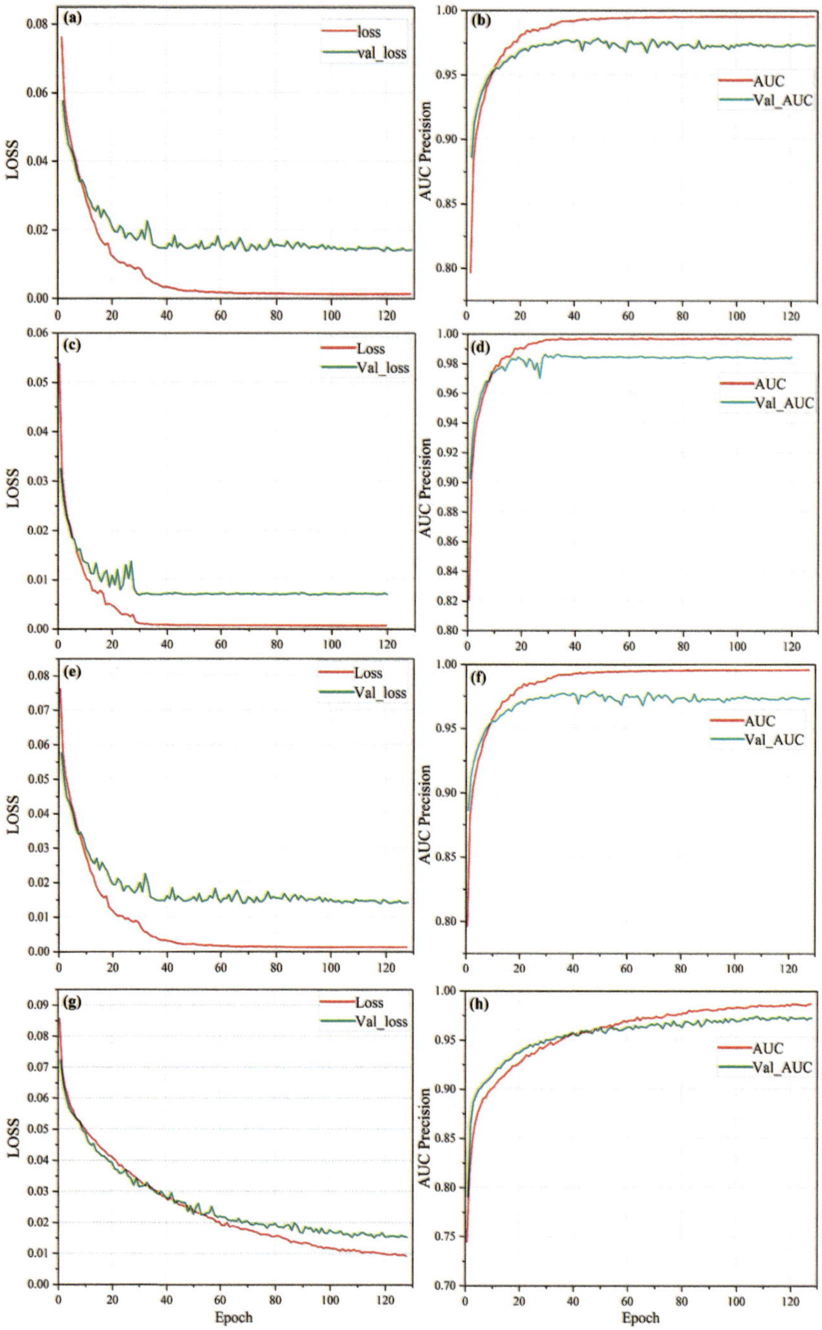

Fig. 6.11 Model training curves: **a** denotes CNN model loss function, **b** denotes CNN model AUC accuracy, **c** denotes MLP model loss function, **d** denotes MLP model AUC accuracy, **e** denotes GRU model loss function, **f** denotes GRU model AUC accuracy, **g** denotes Stacking integrated learning model loss function, **h** denotes Stacking integrated learning model AUC accuracy, reprinted from Ref. Gao et al. (2023), copyright 2025, with permission from IEEE

positive samples correctly predicted as positive by the model, providing a robust indicator of the model's ability to identify landslides.

During the model accuracy evaluation process, the test set is selected as the testing data because it was not involved in the model's training. As the test set is unfamiliar to the model, it more effectively reflects the model's performance and generalization capability. As shown in Fig. 6.12, the Stacking ensemble learning model demonstrates superior performance across all evaluation metrics, with Specificity (0.982), Recall (0.999), MCC (0.982), F1 score (0.991), Kappa (0.982), and OA (0.991) all outperforming those of other models. Among the individual models, the CNN model exhibits the best performance, followed by the MLP, GRU, and SVR models. Notably, the SVR model, used for comparative validation, shows the weakest overall performance. Regarding the Recall metric, the proposed Stacking ensemble learning model achieves the highest Recall (0.999), followed by the MLP model (0.961), CNN model (0.936), SVR model (0.914), and the GRU model, which has the lowest value (0.863). These results, derived from multiple quantitative evaluation metrics, indicate that the Stacking ensemble learning model offers higher overall accuracy in landslide susceptibility evaluation, resulting in more reliable landslide susceptibility maps.

Model uncertainty is a critical aspect of evaluating model performance (Yi et al., 2022). Lv et al. (2022). discussed uncertainty by comparing the average of susceptibility prediction results across various models. This study measures the uncertainty

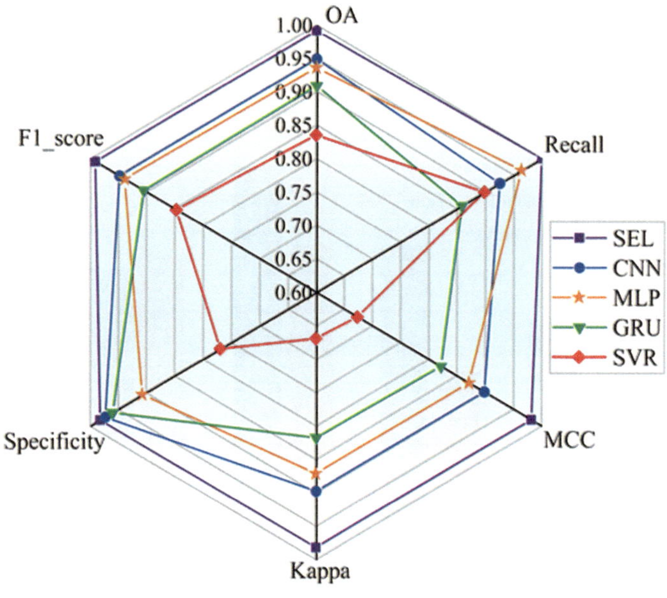

Fig. 6.12 Results of accuracy evaluation of the five models on the test set using the test dataset, reprinted from Ref. Gao et al. (2023), copyright 2025, with permission from IEEE

in model predictions using both the mean and standard deviation. The mean represents the central tendency of the data, while the standard deviation reflects the degree of dispersion.

The standard deviation quantifies the extent of deviation from the mean, with larger values indicating greater deviation and smaller values indicating less deviation. In the model uncertainty analysis shown in Fig. 6.13, the Stacking ensemble learning model developed in this study demonstrates smaller average and standard deviation values (0.17, 0.31) compared to other models. Specifically, the CNN model shows (0.23, 0.41), the MLP model shows (0.41, 0.52), the GRU model shows (0.40, 0.38), and the SVR model shows (0.25, 0.39). These results indicate that the Stacking ensemble learning model exhibits a more concentrated prediction trend, with smaller deviations from the mean, lower uncertainty, higher prediction quality, and superior performance.

After model stabilization, errors between the actual and predicted values were calculated using the test dataset, and error plots were generated (Fig. 6.14). The error plots represent the extent of deviation between predicted and actual values. Smaller deviations correspond to lower errors. Figure 6.14 (a–e) sequentially show the error between actual and predicted values for the Stacking ensemble learning model, CNN model, MLP model, GRU model, and SVR machine learning model on the test set. As shown in the error plot, the Stacking ensemble learning model (Fig. 6.14a) outperforms traditional single models with the lowest MSE and RMSE values (0.15, 0.39), indicating minimal deviation from the true values. The MLP model (Fig. 6.14c) follows with MSE and RMSE values of (0.18, 0.43), while the CNN model (Fig. 6.14b) has MSE and RMSE values of (0.19, 0.44). The GRU model

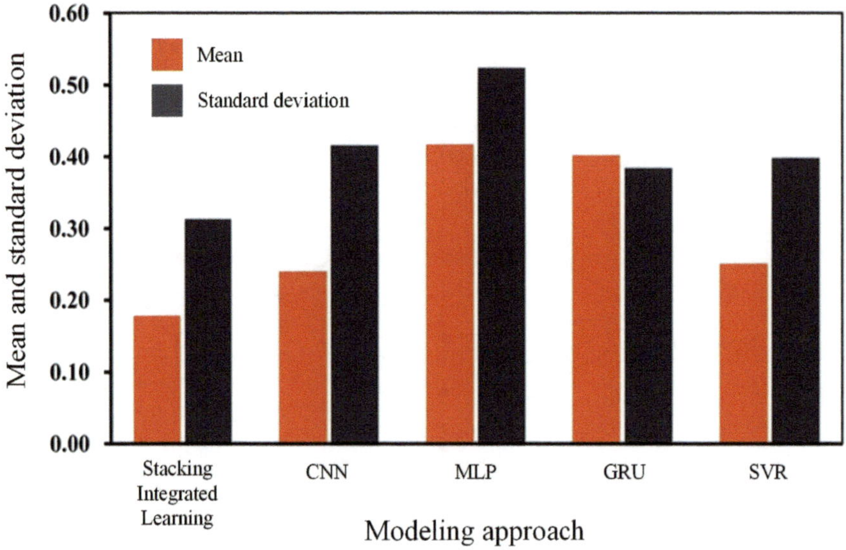

Fig. 6.13 Model uncertainty analysis

(Fig. 6.14d) exhibits MSE and RMSE values of (0.21, 0.46), with a more dispersed error distribution. The SVR model (Fig. 6.14e) shows the largest deviation, with MSE and RMSE values of (0.22, 0.53), and a similarly dispersed error distribution.

The ROC curve is a graphical tool commonly used to assess the performance of binary classification models. It illustrates the trade-off between the true positive rate and the false positive rate at various thresholds. To further evaluate model performance and the reliability of prediction results, this study employs the ROC curve to assess the accuracy and reliability of the Stacking ensemble learning model as well as individual base models (Rong et al., 2023). As depicted in Fig. 6.15, the Stacking ensemble learning model demonstrates the largest area under the ROC curve (AUC), with a value of 0.99, followed by the CNN model (0.98), GRU model (0.98), and MLP model (0.97), while the SVR model exhibits the smallest AUC. Compared

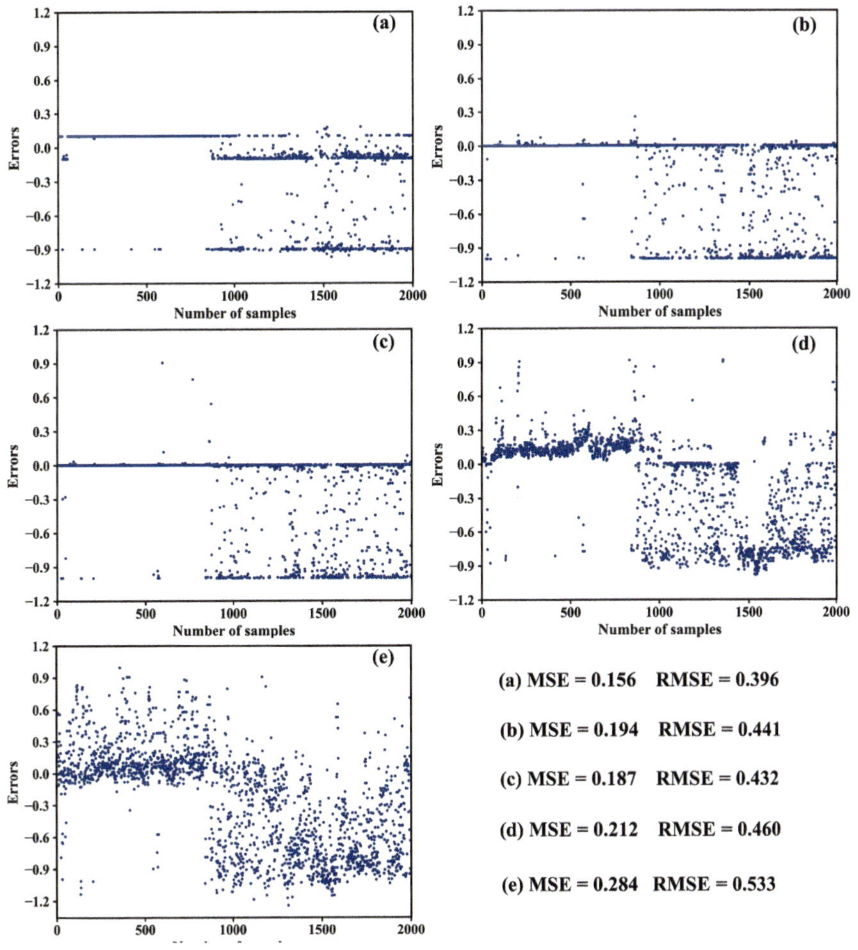

Fig. 6.14 Error analysis of the true and predicted values of the five models using the test dataset

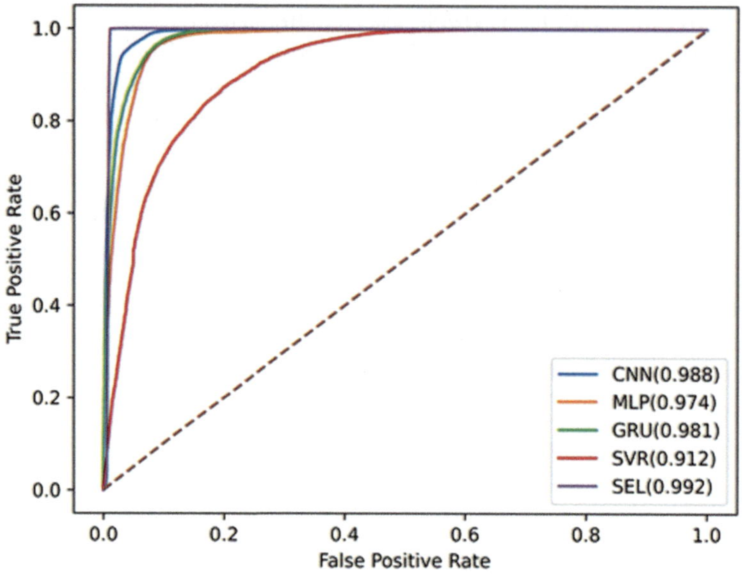

Fig. 6.15 ROC curves for true and predicted values for the five models using the test dataset

to the SVR model, the Stacking ensemble learning model improves accuracy by approximately 9.4%. Based on the combined evaluation from the Recall, Specificity, MCC, OA, Kappa, F1 score, and ROC AUC metrics, it is concluded that the ensemble learning model exhibits superior accuracy and reliability compared to the individual classification models.

6.4.2 Landslide Susceptibility Mapping

Landslide susceptibility mapping plays a critical role in land resource management and environmental protection (Reichenbach et al., 2018). Achieving more accurate landslide susceptibility mapping is essential for mitigating disaster risks and enhancing the scientific basis of land use planning (Fang et al., 2021). Although traditional single models, such as CNN, MLP, and GRU, demonstrate excellent performance in landslide susceptibility mapping and evaluation, they also expose certain limitations when dealing with large-scale remote sensing imagery and complex terrain scenarios, including overfitting and information redundancy. Consequently, this study introduces the Stacking ensemble learning technique for landslide susceptibility mapping, aiming to overcome the limitations inherent in traditional single models. By comparing the Stacking model with conventional models (including CNN, MLP, GRU, and SVR), a more comprehensive assessment of the performance and applicability of various methods in landslide susceptibility prediction is achieved.

Figure 6.16 presents the landslide susceptibility map generated by the Stacking ensemble learning model. The areas with the highest landslide susceptibility are primarily concentrated in the central region of Shenzhen, characterized by mountainous terrain with significant elevation changes. Smaller areas of susceptibility are observed along the coastal regions of Yantian District and the northern part of Longgang District. Historical and potential future landslides are predominantly located in the extremely high, high, and moderate susceptibility zones. Extremely low susceptibility areas are mainly found in low-altitude and low-slope regions. The transitional effects between different susceptibility levels exhibit a smooth trend, with susceptibility being strongest at the landslide center and gradually decreasing outward. This pattern is consistent with field observations of the influence of landslides on the surrounding environment. As the distance from the center of the landslide body increases, the impact strength diminishes, leading to a lower probability of landslide occurrence and a corresponding reduction in susceptibility (Liu et al., 2024a, 2024b). This further validates the reasonableness and consistency of the landslide susceptibility map generated by the Stacking ensemble learning model in this study with actual field observations.

To further assess the effectiveness of the ensemble learning model developed in this study, a series of ablation experiments were conducted using individual classifier models (CNN, MLP, GRU, and SVR). The landslide susceptibility mapping for each single classifier model are presented in Fig. 6.17. Specifically, Fig. 6.17a, b, c, and d show the landslide susceptibility maps generated using the CNN, MLP, GRU, and SVR models, respectively. All individual neural network classifier models yielded

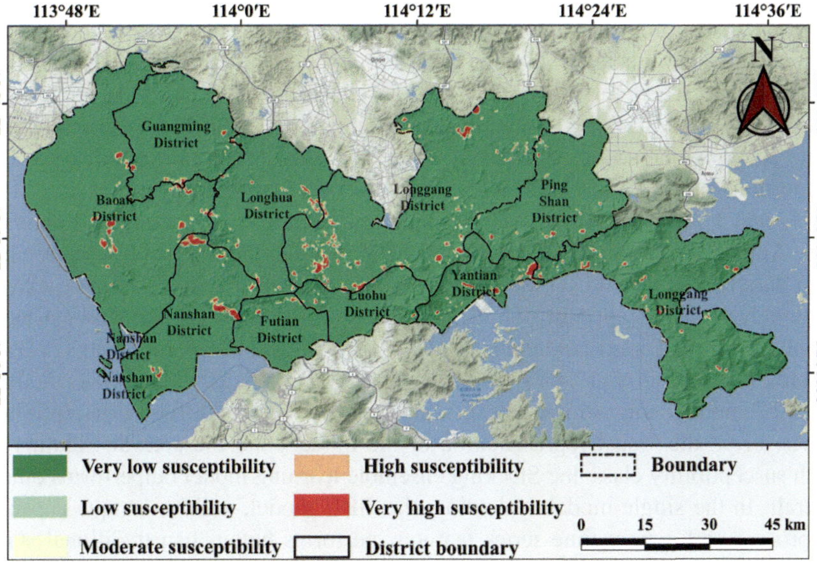

Fig. 6.16 Stacking integrated learning model landslide susceptibility map, reprinted from Ref. Gao et al. (2023), copyright 2025, with permission from IEEE

relatively accurate landslide susceptibility maps. Visually, areas with a high likelihood of landslides overlap with historical and potential landslide points. Notably, the landslide susceptibility maps from the CNN and GRU models resemble those produced by the Stacking ensemble learning model. However, the CNN model's susceptibility map displays poor transitions between different susceptibility zones. This could be attributed to the strong feature extraction capabilities of the convolutional layers in the CNN model (Xia et al., 2021), leading to extreme distributions in the predicted susceptibility values, with distinct delineation between landslide and non-landslide regions. In contrast, the GRU model, by incorporating time-series dynamic features through gating units in its recurrent layers, enhances the model's performance in landslide susceptibility evaluation. The MLP model predicts an extensive distribution of the extremely high and high susceptibility areas, spanning the central region of Shenzhen. In comparison, the landslide susceptibility map produced by the SVR model is less reliable than those generated by deep learning models, and it does not align well with actual observations. Furthermore, in regions with relatively mild terrain and gentle slopes, the landslide susceptibility maps generated by the MLP (Fig. 6.17b), GRU (Fig. 6.17c), and SVR (Fig. 6.17d) models show higher and moderate susceptibility. This phenomenon may be attributed to the models' tendency to overfit, resulting in reduced generalization ability and an inability to effectively capture the features of areas with subtle terrain undulations and low slopes. In such cases, the models may overly rely on noise or local patterns in the training data, disregarding the broader terrain regularities.

Figure 6.18 illustrates the area distribution derived from landslide susceptibility maps obtained using the Stacking ensemble learning model and individual models (CNN, MLP, GRU, and SVR), calculated through statistical methods based on different susceptibility levels. Figure 6.18a, b, c, d display the susceptibility maps generated by the Stacking ensemble learning model, CNN, MLP, GRU, and SVR models, respectively. As shown in Fig. 6.18a, the majority of the landslide-prone areas generated by the Stacking ensemble learning model are classified as low susceptibility zones, which occupy a relatively large proportion of the entire study area. The proportions of extremely high, high, and moderate susceptibility zones are 4.1%, 7.8%, and 1%, respectively, with the high susceptibility area being the dominant one. Among the CNN, MLP, GRU, and SVR models, the landslide susceptibility map derived from the GRU model (Fig. 6.18d) exhibits a relatively small area of extremely high susceptibility (4.5%). In contrast, the SVR model shows a more evenly distributed area of extremely high (10.6%), high (11.1%), moderate (17.3%), and low susceptibility (12.4%) zones. In comparison to the CNN and GRU models, the MLP model demonstrates a better transition between the different susceptibility zones. From the visual representation of the models and the area distribution of each susceptibility class, the Stacking ensemble learning model outperforms others overall. In the single model evaluation, the GRU model, with its unique capacity to process and extract time-series features, performs better than traditional static models when addressing long-term dynamic landslide impact factors. Unlike single classifiers such as CNN, GRU, MLP, and SVR, the Stacking ensemble learning model leverages the strengths of each individual model in data handling, resulting

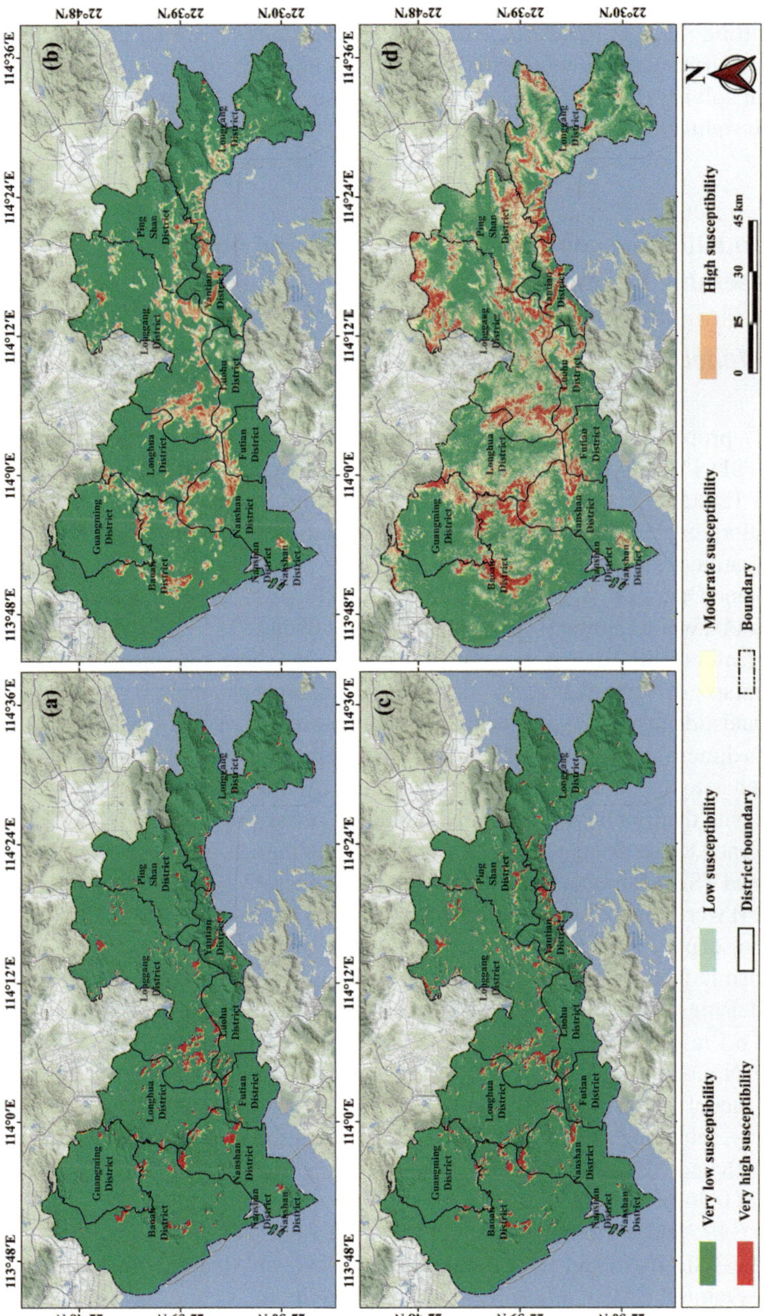

Fig. 6.17 Single-model landslide susceptibility map, reprinted from ref.Ref. Gao et al. (2023), copyright 2025, with permission from IEEE

in superior feature extraction capabilities. The Stacking ensemble learning model constructed in this study successfully addresses the challenges of poor generalization and overfitting inherent in single models, delivering superior overall performance. This result further confirms the effectiveness and potential of Stacking ensemble learning in solving complex problems and enhancing model performance, offering valuable insights for related research and practical applications.

6.5 Landslide Susceptibility Mapping Based on the TSDNN Ensemble Framework

6.5.1 Model Performance Analysis

This study proposes a multi-network integrated landslide susceptibility mapping method (TSDNN) that incorporates both static and dynamic features. The method integrates 12 static landslide impact factors and 2 time-series dynamic landslide impact factors, enabling parallel extraction of both static spatial features and temporal dynamic features of landslides. In the data processing phase, the dataset is divided into Dataset A and Dataset B. Specifically, Dataset A includes 12 selected static landslide impact factors, while Dataset B comprises two types of data: Type 1 contains 12 static landslide factors with a data dimension of (13, 13, 12), where 13 represents the row and column size of each patch, and 12 denotes the number of channels, i.e., the number of static landslide factors. Type 2 consists of 2 time-series dynamic landslide factors with a data dimension of (48, 13, 13, 2), where 48 represents the time dimension, and 2 corresponds to the number of channels, i.e., the two time-series dynamic factors.

To evaluate the training accuracy and reliability of the proposed TSDNN model, Dataset A and Dataset B were split into training and testing sets in a 7:3 ratio. Based on the proposed TSDNN model, a series of rigorous comparative experiments (Girshick et al., 2014) were conducted using Dataset A, and the accuracy and reliability of the landslide susceptibility evaluation were further validated by incorporating the time-series dynamic factors from Dataset B. The training parameters of all models were optimized using grid search (Xiang et al., 2023), as detailed in Table 6.5.

Figure 6.19a and b present the variation curves for LogCosh and AUC values of the MSCNN model based on Dataset A. On the training set, after 60 iterations, the MSCNN model stabilized with LogCosh and AUC values of 0.0011 and 0.9952, respectively, achieving an F1 value of 0.9974. On the testing set, after 100 iterations, the LogCosh and AUC values stabilized at 0.0143 and 0.9732, respectively, with an F1 value of 0.9628 (Table 6.6).

Figure 6.19c and d show the variation curves of LogCosh and AUC values after training the TSDNN model with Dataset B. We observed that the introduction of time-series rainfall data significantly accelerated the model's convergence. After 30 iterations, the accuracy curve exhibited greater stability than that of the MSCNN model. When the TSDNN model stabilized, the LogCosh value on the training

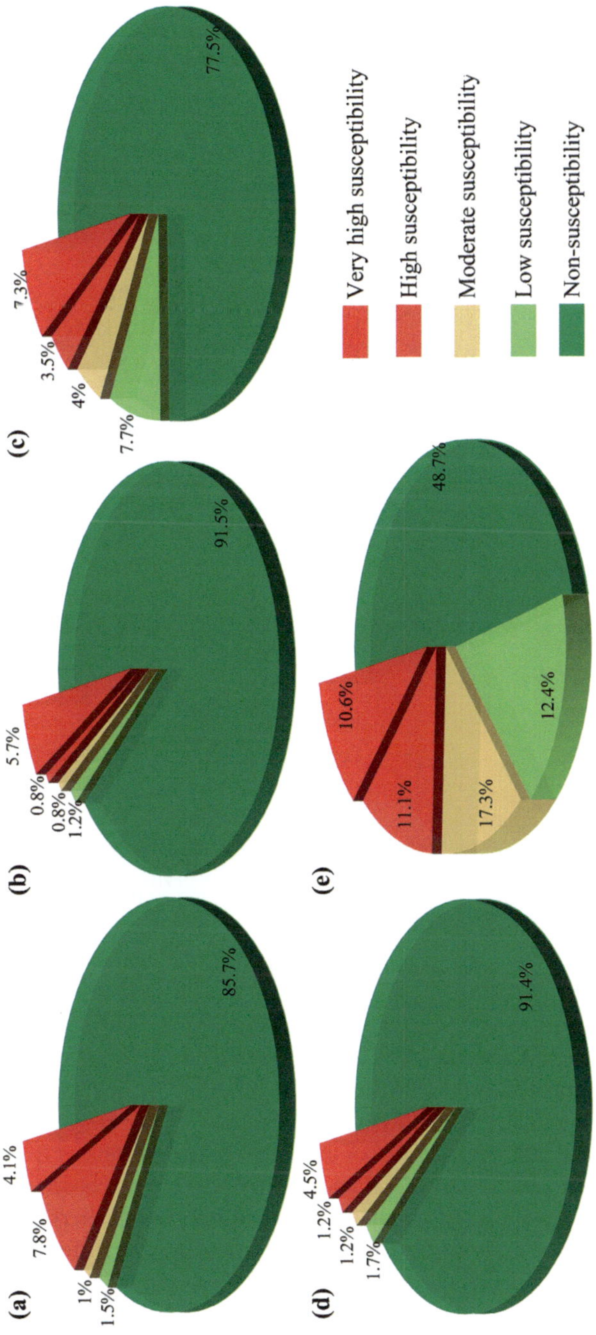

Fig. 6.18 Stacking integrated learning model landslide susceptibility area statistics

Table 6.5 Detailed parameters for model training

Model name	Input dataset	Parameter name	Optimal parameters
MSCNN model	Dataset A	Batch size	128
		Epochs	128
		Learning rate	0.00000640
		Dropout	0.2
		Kernal size	3
		Factor	0.75
TSDNN model	Dataset B	Batch size	64
		Epochs	128
		Learning rate	0.00004
		Dropout	0.5
		Kernal size	3
		Factor	0.1
MSCNN model		Batch size	128
		Epochs	128
		Learning rate	0.00000134
		Dropout	0.2
		Kernal size	3
		Factor	0.7
CNN model		Batch size	128
		Epochs	128
		Learning rate	0.000001
		Dropout	0.3
		Kernal size	3
		Factor	0.75
RF model		n_estimators	100
		Criterion	gini
		max_depth	6

set was lower (0.0006) and the AUC value was higher (0.9964). On the testing set, the TSDNN model showed consistent LogCosh (0.007), AUC (0.9843), Recall (0.99873), Precision (0.9542), and F1 (0.9705) values.

To further evaluate the performance of the TSDNN model, Dataset B was divided using the same ratio, and the MSCNN and CNN models were chosen as baselines for comparison. To control for variables, identical hyperparameters and iteration numbers were used in the comparative experiments, with the only difference being the input data. Figure 6.19e-h depict the changes in LogCosh and AUC indicators during iterative training for both models. The MSCNN model (Fig. 6.19e, f) performed well on the training set and stabilized after 100 iterations on the testing set. The CNN

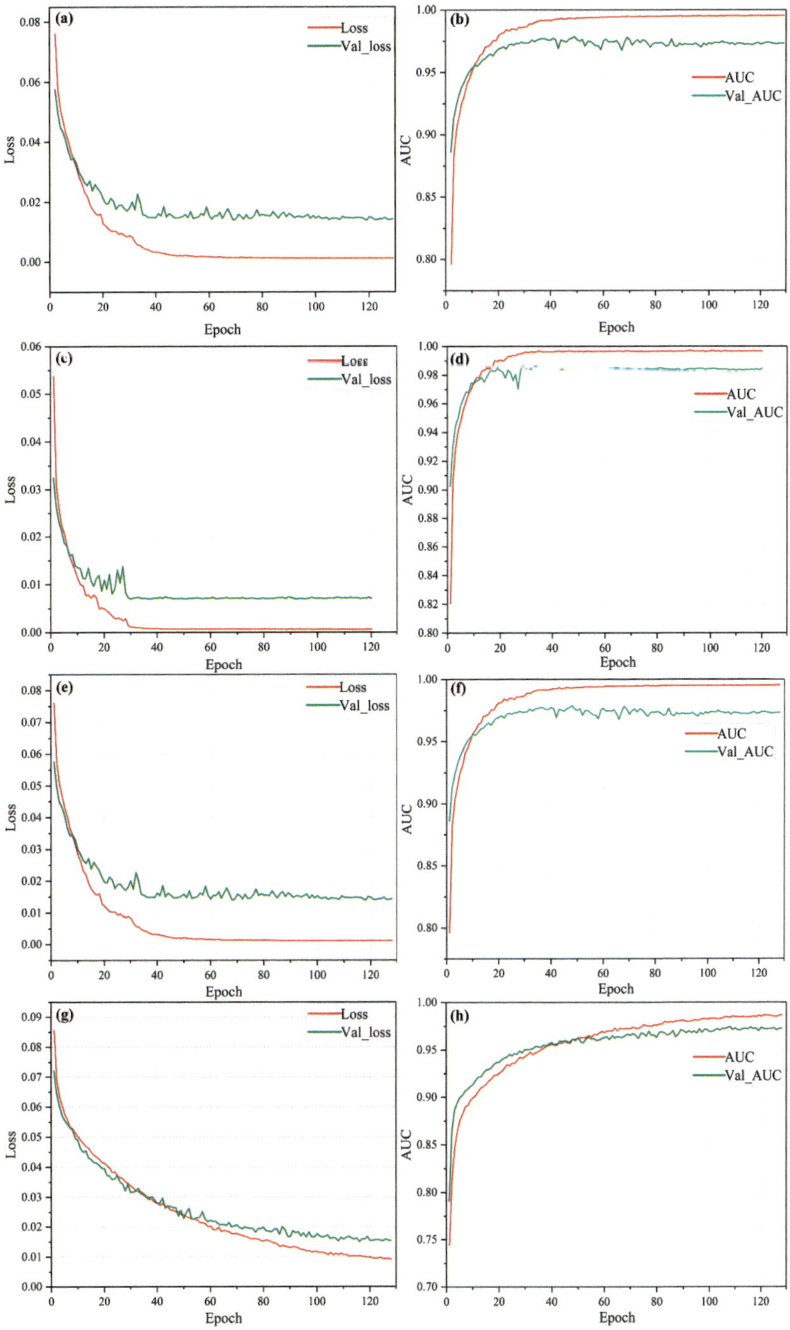

Fig. 6.19 Model training curves: **a** logcosh of the MSCNN model using dataset A, **b** AUC of the MSCNN model using dataset A, **c** logcosh of the MSCNN model using dataset B, **d** AUC of the TSDNN model using dataset B, **e** MSCNN model using dataset B logcosh, **f** AUC using the MSCNN model of dataset B, **g** logcosh using the CNN model of dataset B, and **h** AUC using the CNN model of dataset B, reprinted from Ref. Gao et al. (2024), with permission from IEEE

Table 6.6 Results of model accuracy evaluation

Dataset	Model name		LogCosh	AUC	Recall	Precision	F1
Dataset A	MSCNN model	Train	0.001	0.995	0.999	0.995	0.997
		Test	0.014	0.973	0.985	0.940	0.962
Dataset B	TSDNN model	Train	0.000	0.996	0.999	0.996	0.998
		Test	0.007	0.984	0.987	0.954	0.970
	MSCNN model	Train	0.001	0.994	0.992	0.998	0.995
		Test	0.006	0.978	0.986	0.966	0.976
	CNN model	Train	0.013	0.974	0.955	0.993	0.943
		Test	0.022	0.950	0.998	0.984	0.910
	RF model	Train	–	0.951	0.935	0.819	0.873
		Test	–	0.921	0.683	0.615	0.647

model (Fig. 6.19g, h) gradually stabilized after 120 iterations. Table 6.6 presents the performance evaluation metrics after 120 iterations. In terms of the LogCosh indicator, the values were: TSDNN (0.0006) < MSCNN (0.0012) < CNN (0.0135). For AUC, Recall, and F1 indicators, the values were: TSDNN (0.996, 0.999, 0.998) > MSCNN (0.994, 0.992, 0.995) > CNN (0.974, 0.955, 0.955). For the precision indicator, the values were: MSCNN (0.998) > CNN (0.993) > TSDNN (0.996).

Overall, the evaluation metrics demonstrate that while the MSCNN model exhibits high learning ability and reliability when using Dataset A, the proposed TSDNN model shows more stable overall performance across all evaluation indicators when time-series rainfall data is introduced. Specifically, the LogCosh converges faster, and the AUC, Recall, Precision, and F1 indicators are optimized. These findings suggest that incorporating time-series rainfall data into traditional static factors can significantly enhance the accuracy of landslide susceptibility mapping.

This study introduces the integration of time-series dynamic data with static landslide influencing factors, highlighting that the inclusion of time-series dynamic data can enhance the accuracy and reliability of landslide susceptibility mapping. To evaluate the model's performance in processing time-series dynamic data, we conduct a comparative analysis focusing on model uncertainty, error maps, and ROC curves.

As shown in Fig. 6.20, the model uncertainty analysis reveals that the TSDNN model proposed in this study has both lower mean and standard deviation values (0.05, 0.20) compared to the MSCNN model (0.06, 0.23). This indicates that the prediction results of the TSDNN model are more concentrated, with a smaller degree of deviation from the mean, lower uncertainty, and higher prediction quality, thus demonstrating superior model performance. In contrast, the CNN model (0.29, 0.41) shows the highest mean and standard deviation, suggesting a lack of clear central tendency in its predictions, with higher dispersion and greater uncertainty.

Once the model training reached stability, the errors between the actual and predicted values were computed using the test dataset, and error maps for the models were generated (Fig. 6.21). Figure 6.21a and b display the error maps for

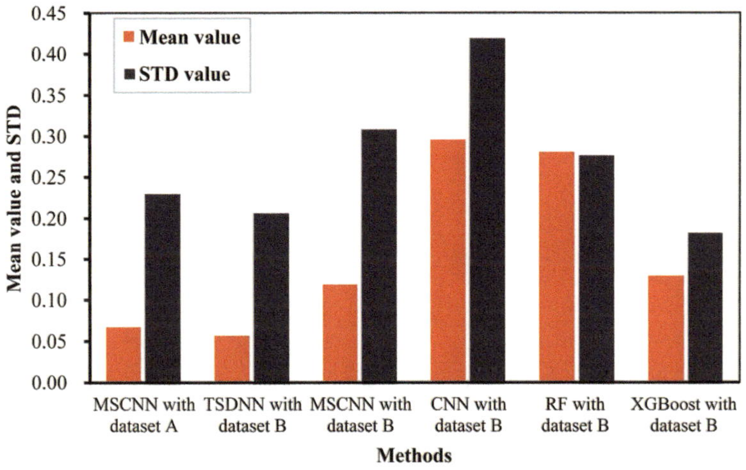

Fig. 6.20 Model uncertainty analysis, reprinted from Ref. Gao et al. (2024), with permission from IEEE

the MSCNN and TSDNN models on the test set, using static dataset A and static-dynamic dataset B, respectively. Figure 21b–f depict the error maps for the TSDNN, MSCNN, CNN, RF, and Extreme Gradient Boosting (XGBoost) models when using only static-dynamic dataset B. From the error maps, it is evident that the TSDNN model (Fig. 21b) yields smaller MSE and RMSE values (0.20, 0.45) in both dataset A and dataset B compared to the MSCNN model (0.21, 0.41) (Fig. 21a), indicating that the TSDNN model offers the smallest deviation from the true values. When processing static-dynamic dataset B, the TSDNN model exhibits the smallest errors, followed by the CNN model (Fig. 21d), with MSE and RMSE values of (0.26, 0.51) and a more scattered error distribution. The RF model (Fig. 21e) demonstrates larger deviations from the true values, with an even more dispersed error distribution.

ROC (Peng et al., 2015) is frequently used to characterize the relationship between a model's predicted values and the true values, offering a comprehensive evaluation of model performance. As shown in Fig. 6.22, the TSDNN model achieves the highest ROC value 0.99, among all the models compared, followed by the MSCNN model (0.96). These findings suggest that incorporating time-series rainfall landslide factors can significantly enhance the accuracy of susceptibility mapping. When using dataset B, the MSCNN model scores 0.88, the RF model scores 0.88, and the CNN model scores 0.86, all of which are considerably lower than the TSDNN model. Therefore, the TSDNN model proposed in this study is recommended as the best model for landslide susceptibility mapping.

Further, the interpretability of the model is analyzed. As is well known, ML/DL models have been widely applied in the field of Earth sciences and are often referred to as black-box models. Therefore, the interpretability of a model can provide researchers and decision-makers with an intuitive and meaningful explanation of the model (Adadi & Berrada, 2018).

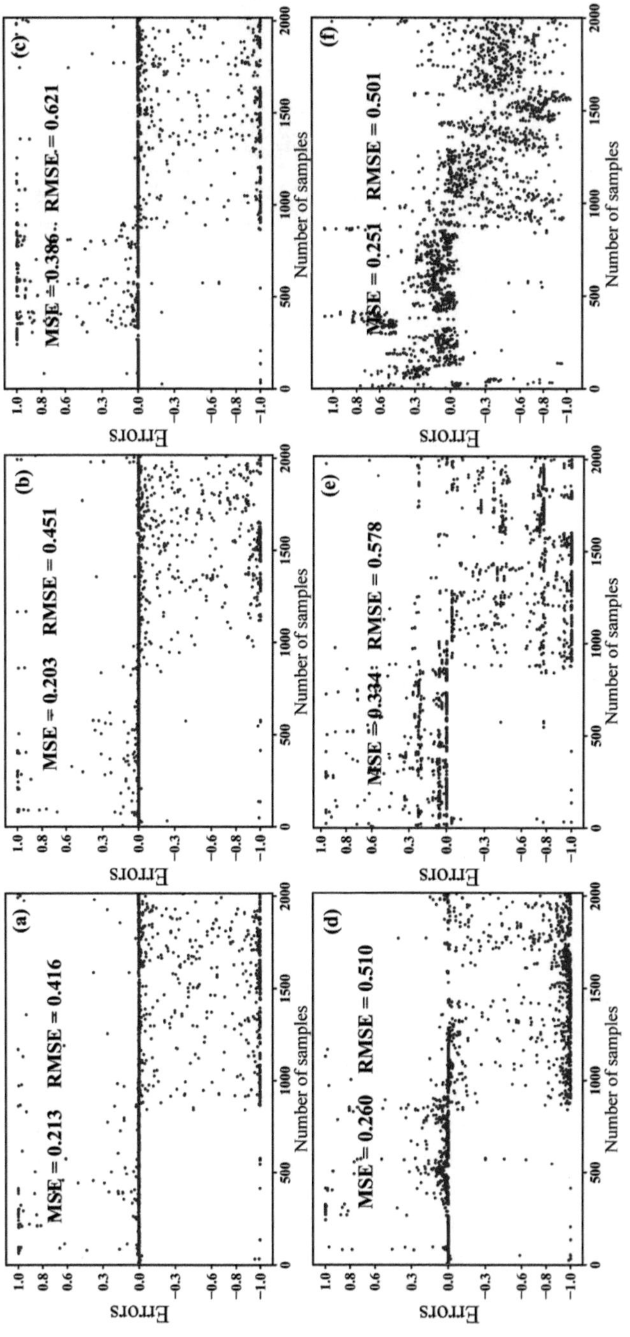

Fig. 6.21 Error analysis of the true and predicted values of the six models using the test dataset, reprinted from Ref. Gao et al. (2024), copyright 2025, with permission from IEEE

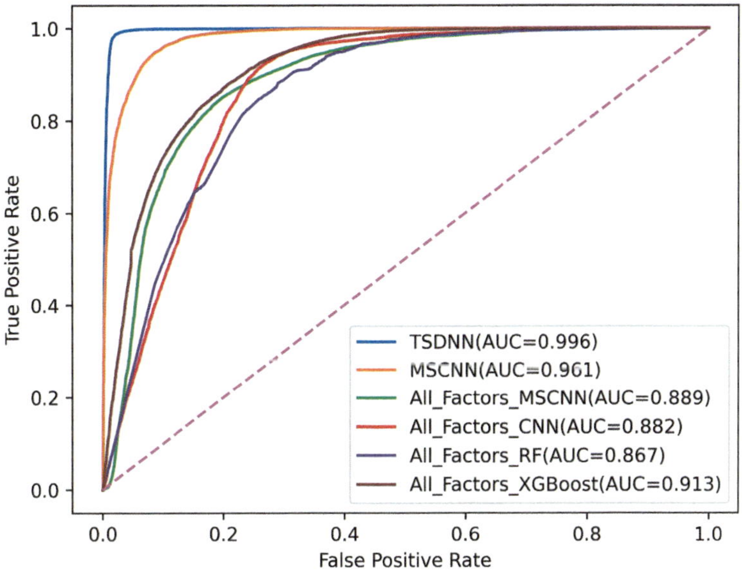

Fig. 6.22 ROC curves for true and predicted values for the six models using the test dataset, reprinted from Ref. Gao et al. (2024), copyright 2025, with permission from IEEE

SHAP is regarded as an alternative to traditional feature importance methods (Al-Najjar et al., 2023). It provides a visual representation to demonstrate how features influence the dependent variable and thoroughly explains the model components (Kim et al., 2022). The SHAP model is a more comprehensive approach for model interpretability, enabling both global and local interpretations. By calculating the incremental contribution of a feature when added to the model and considering the incremental contributions across all feature permutations, an average value is derived, which represents the SHAP baseline value for that feature.

The SHAP method is applied to analyze the interpretability of the proposed TSDNN model. A random selection approach is employed to select time-series rainfall samples and analyze the contribution of dynamic features within the TSDNN model. The SHAP heatmap for the interpretation of time-series features is presented in this study (Fig. 6.23). In Fig. 6.23, the grayscale image represents the time-series rainfall samples at each time point, while the color heatmap, ranging from blue to red, indicates the extent to which the features explain the landslide. Red indicates a positive contribution to the landslide, while blue indicates a negative contribution. In this case, red signifies a positive impact on the landslide, and blue denotes a negative impact. It is evident that in the 48 time-series rainfall feature maps, the proportion of red is relatively high. Therefore, the time-series rainfall features used in this study demonstrate strong explanatory power for landslides, significantly enhancing model performance and contributing to a better understanding of landslides.

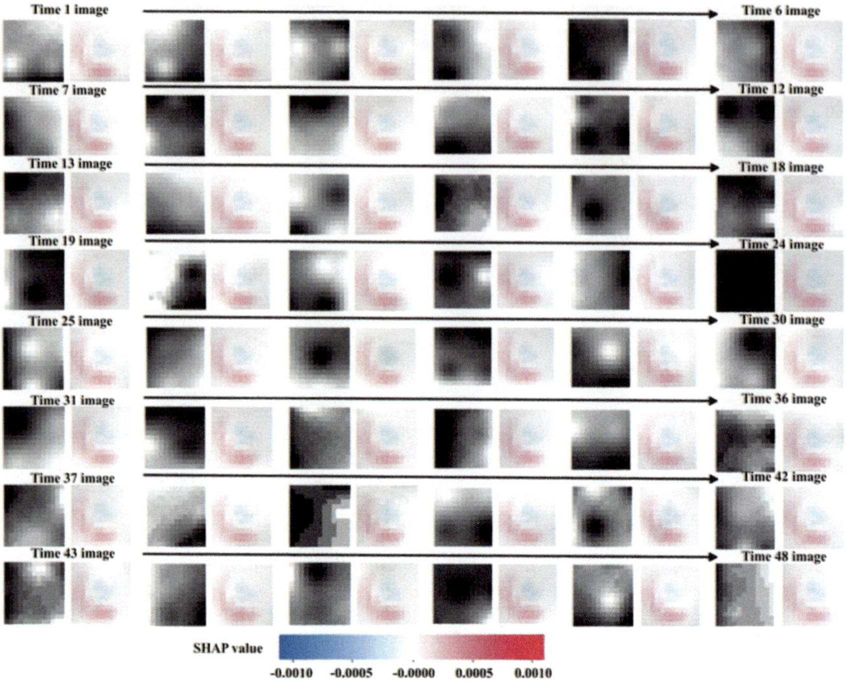

Fig. 6.23 Images of SHAP values for the TSDNN model, reprinted from Ref. Gao et al. (2024), copyright 2025, with permission from IEEE

6.5.2 Landslide Susceptibility Mapping

This study builds upon prior research (Gao et al., 2023) and focuses on landslide susceptibility mapping that accounts for both static and dynamic influencing factors. To more effectively incorporate time-series dynamic landslide factors, we propose a multi-network ensemble method, TSDNN, for landslide susceptibility mapping, capable of processing both static and dynamic factors simultaneously. This method integrates 12 selected static landslide influencing factors and 2 time-series dynamic landslide factors. After ensuring stable model training and reliable accuracy evaluation, we generate a predicted dataset using pixel-wise sliding clipping, which is then input into the proposed TSDNN model, along with comparison models such as MSCNN, CNN, RF, and XGBoost, to generate landslide susceptibility maps (Fig. 6.24).

Specifically, the predicted dataset A, containing only the 12 static landslide influencing factors, is input into the MSCNN model (Fig. 6.24a). The predicted dataset B, which includes both the 12 static factors and 2 time-series dynamic landslide influencing factors, is input into the TSDNN (Fig. 6.24b), MSCNN (Fig. 6.24c), CNN (Fig. 6.24d), RF (Fig. 6.24e), and XGBoost models (Fig. 6.24f) to generate landslide

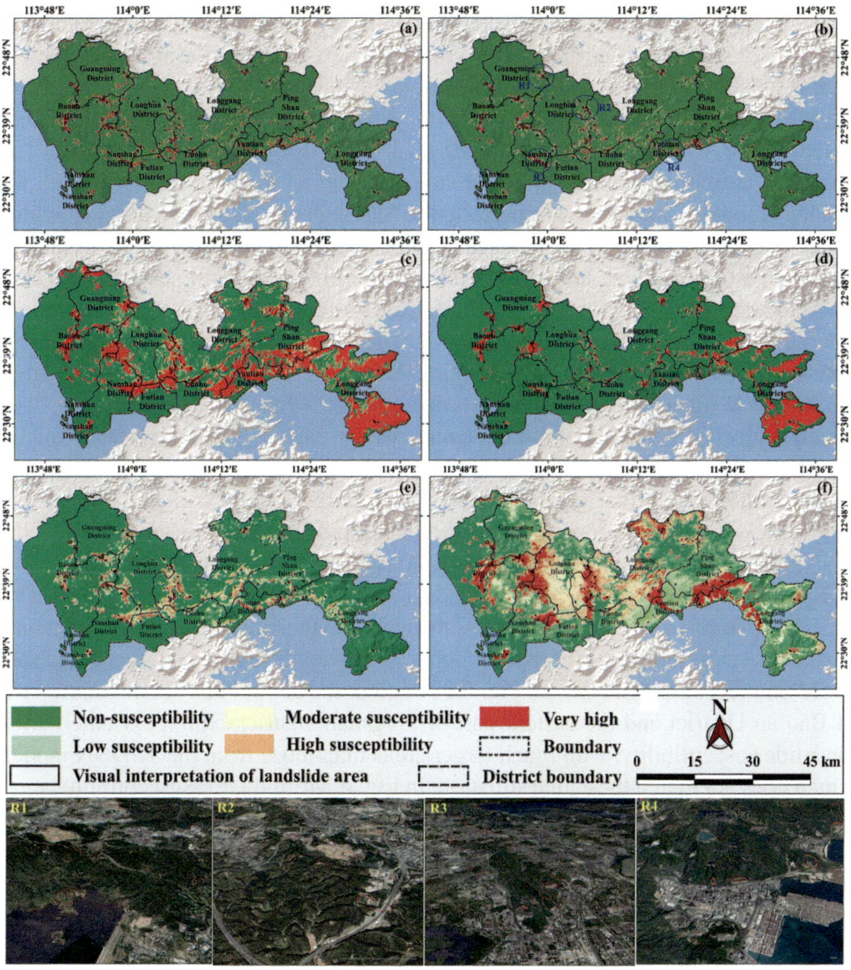

Fig. 6.24 Landslide susceptibility mapping results **a** MSCNN model using dataset A, **b** TSDNN model using dataset B, **c** MSCNN model using dataset B, **d** CNN model using dataset B, **e** RF model using dataset B, **f** XGBoost model using dataset B. R1, R2, R3 and R4 are the remote sensing images of some susceptibility areas predicted by the TSDNN model, reprinted from Ref. Gao et al. (2024), copyright 2025, with permission from IEEE

susceptibility maps. The MSCNN model results based on dataset A serve as an ablation experiment, assessing the impact of adding time-series dynamic factors to the landslide susceptibility mapping. With dataset B, several popular neural network and machine learning models are compared to evaluate the performance and potential of each model in handling both static and dynamic landslide influencing factors.

The results of the landslide susceptibility mapping considering both static and dynamic influencing factors show that, compared to maps generated using only dataset A, those based on dataset B are generally better. The regions with extremely high susceptibility to landslides overlap with historical landslides, and more importantly, these regions are further refined (Fig. 6.24a, b). Visually, the area of extremely high susceptibility based on the TSDNN model using dataset B is smaller than that of the MSCNN model using dataset A, but the high and moderate susceptibility areas based on the TSDNN model are larger than those from the MSCNN model. The transition between high and moderate susceptibility areas using dataset B with the TSDNN model is smoother compared to the MSCNN model using dataset A, which better matches the real-world influence of landslides on the surrounding environment, making it more practical. It is noteworthy that after introducing time-series dynamic landslide influencing factors, the moderate susceptibility areas in the central-western Bao'an District and western Guangming District, which are smaller in area, are less than those from the MSCNN model using dataset A. These areas may result from the MSCNN model's over-reliance on noise or local features in the training data, which neglects the overall topographic regularities, causing the model to incorrectly identify noise regions as landslide zones.

To further verify the advantages of the proposed TSDNN model in handling time-series dynamic factors, the landslide susceptibility maps from the MSCNN, CNN, RF, and XGBoost models using dataset B are compared (Fig. 6.24b, c, d, e, f). Specifically, the TSDNN model demonstrates the best performance, followed by MSCNN, while the RF model performs the worst. In these comparative experiments, the areas of extremely high, high, and moderate susceptibility are all larger than those based on the TSDNN model. In Fig. 6.24c and d, the central and southeastern parts of Bao'an District and the eastern part of Pingshan District exhibit extremely high landslide susceptibility, with much larger areas than those from the TSDNN model, which does not match the actual situation. In Fig. 6.24e and f, the susceptibility areas generated by the RF and XGBoost models using dataset B are significantly larger than those from the TSDNN model (Fig. 6.24b). The RF model results suggest moderate landslide susceptibility in the central urban area of Shenzhen, which is inconsistent with reality. This discrepancy may arise from the MSCNN, CNN, RF, and XGBoost models' insufficient ability to handle time-series dynamic factors, leading to data redundancy and overfitting. In contrast, the TSDNN model excels at extracting both spatial and time-series dynamic features. This is due to its unique parallel architecture combining MSCNN and Bi-LSTM. The MSCNN extracts static landslide features at various depths, while the TD-CNN structure converts the 2D spatial domain features of time-series dynamic factors into 1D sequence features, which are then processed by the Bi-LSTM structure to capture dynamic temporal features. Finally, the TSDNN model integrates both spatial and dynamic time-series features, producing more reliable and universally applicable landslide susceptibility maps.

To quantitatively assess the landslide susceptibility mapping results produced by the proposed TSDNN model, the outputs from each model were reclassified using the Natural Breaks method in QGIS, and the proportion of each susceptibility class relative to the total area was calculated through statistical analysis (Fig. 6.25). The landslide susceptibility map generated by the TSDNN model with dataset B (Fig. 25b) was more refined than that of the MSCNN model using dataset A (Fig. 25a), with notably larger areas classified as high and moderate susceptibility. When compared to MSCNN, CNN, RF, and XGBoost models trained on dataset B (Fig. 25c–f), the TSDNN model predicted a smaller proportion of extremely high susceptibility areas (3.9%) than MSCNN (10.3%), CNN (23.7%), RF (14.5%), and XGBoost (4.8%). The moderate-to-high susceptibility areas (1.2% each) were also smaller than those of the CNN model (4.4% and 3.3%), but slightly larger than those of the MSCNN model (1.1% and 1.0%). These findings suggest that, when both static and dynamic landslide conditioning factors are considered, conventional static neural network and machine learning models exhibit lower predictive accuracy. This is likely due to the limitations of two-dimensional convolutional layers, which are effective for static spatial features but struggle to capture temporal dynamics, such as time-series rainfall. Furthermore, the MSCNN and CNN architectures lack temporal processing components, making it difficult to properly integrate dynamic data. When large volumes of time-series data are treated as static inputs, these models may overemphasize such features, leading to overfitting and poor generalization. In contrast, the proposed TSDNN model, which integrates static and dynamic features via a multi-stream architecture, exhibits superior performance and generalization capability, offering more accurate and reliable results in landslide susceptibility mapping.

InSAR technology offers valuable reference data for enhancing the accuracy of landslide susceptibility mapping. In this study, the susceptibility results generated by the proposed TSDNN model were integrated with InSAR-derived deformation measurements to identify and assess potentially hazardous zones exhibiting slow ground movement. The landslide susceptibility mapping outputs were reclassified using the Natural Breaks method in QGIS, and the resulting classes were employed as masks to extract corresponding InSAR deformation data. Subsequently, histograms illustrating the distribution of InSAR deformation values across different susceptibility levels were constructed (Fig. 6.26). As shown in Fig. 6.26a, the deformation intensity is predominantly concentrated to the left of 0.7, with a grid cell density of approximately 15, and the peak of the normal distribution curve located near 0.7. With increasing susceptibility levels, the deformation grid density rises, and the distribution peak gradually shifts to the right. In areas of extremely high susceptibility (Fig. 6.26e), the deformation grid density reaches 18, and the distribution peak occurs near 0.83. These results demonstrate that the deformation intensity across all susceptibility classes approximates a normal distribution and increases with susceptibility level (Bouvry et al., 2011). This trend indicates a positive correlation between landslide susceptibility and deformation intensity, aligning well with observed patterns of landslide occurrence in field investigations (Huang et al., 2022).

To further validate the reliability of the landslide susceptibility predictions generated by the proposed TSDNN model, two representative high-risk areas were selected for case analysis. Historical high-resolution satellite imagery from Google Earth (Dai et al., 2022; Ju et al., 2022), combined with InSAR based ground surface deformation data, was used to assess the model's applicability in real-world scenarios. Region 1 is located north of the Sanlian Reservoir, where anthropogenic modifications to the terrain have led to the development of unstable slopes. As shown in Fig. 6.27a, the TSDNN model predicts an area of extremely high susceptibility, with susceptibility

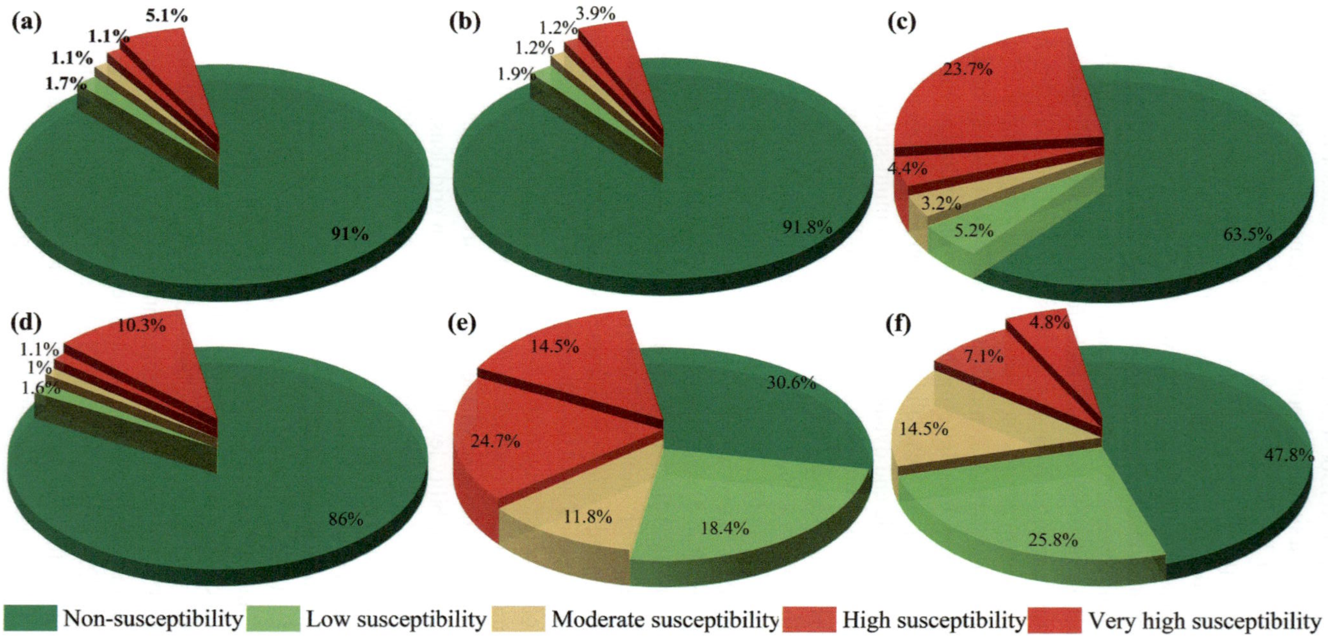

Fig. 6.25 Landslide susceptibility area statistics taking into account static-dynamic landslide impact factors **a** MSCNN model using dataset A, **b** TSDNN model using dataset B, **c** MSCNN model using dataset B, **d** CNN model using dataset B, **e** RF model using dataset B, and (f) XGBoost model using dataset B, reprinted from Ref. Gao et al. (2024), copyright 2025, with permission from IEEE

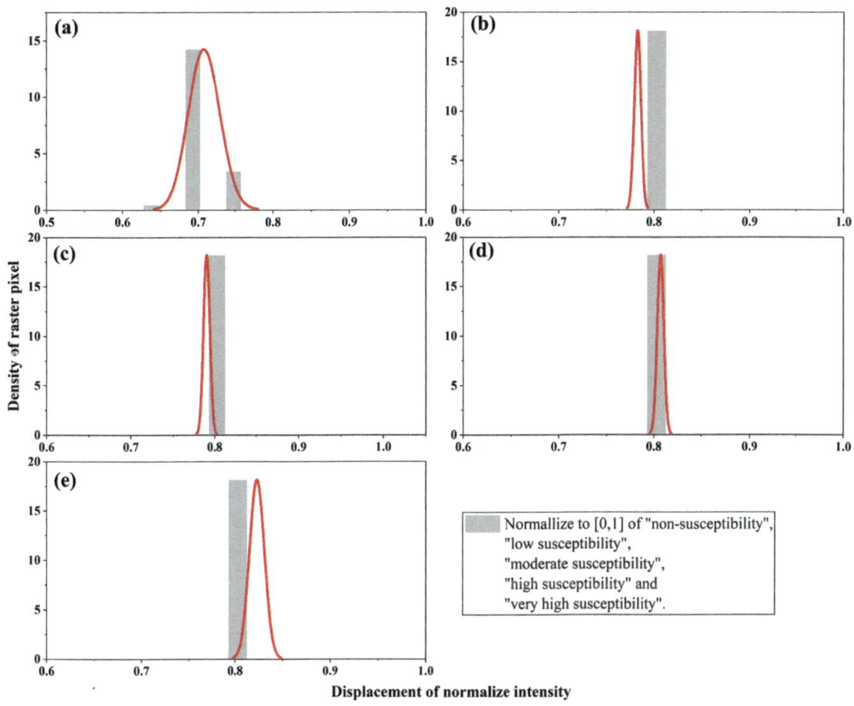

Fig. 6.26 InSAR deformation intensity for different landslide susceptibility **a** very low sensitivity, **b** low sensitivity, **c** medium sensitivity, **d** high sensitivity, **e** very high sensitivity, red curves are normal distribution, reprinted from Ref. Gao et al. (2024), copyright 2025, with permission from IEEE

levels decreasing radially from the center. The InSAR deformation results for this area (Fig. 6.27b) reveal a pattern of gradual ground movement from 2018 to 2021, with a maximum cumulative deformation rate of –14 mm/year, indicating a latent hazard.

For Region 2, the predicted susceptibility levels (Fig. 6.28a) range from extremely high to moderate. The InSAR deformation results (Fig. 6.28b and c) show a pronounced subsidence trend, with deformation rates reached -11 mm/year. Of particular note, the high-susceptibility locations A, B, and C identified by the TSDNN model (Fig. 6.28d) correspond to the sites of major landslide events triggered by heavy rainfall in October 2023, resulting in significant economic losses. Given the area's severe landslide risk, the presence of densely built environments, and the intensity of human activities, continuous real-time monitoring is recommended. The case analyses for Regions 1 and 2 confirm that the TSDNN model offers high accuracy and strong reliability in landslide susceptibility mapping.

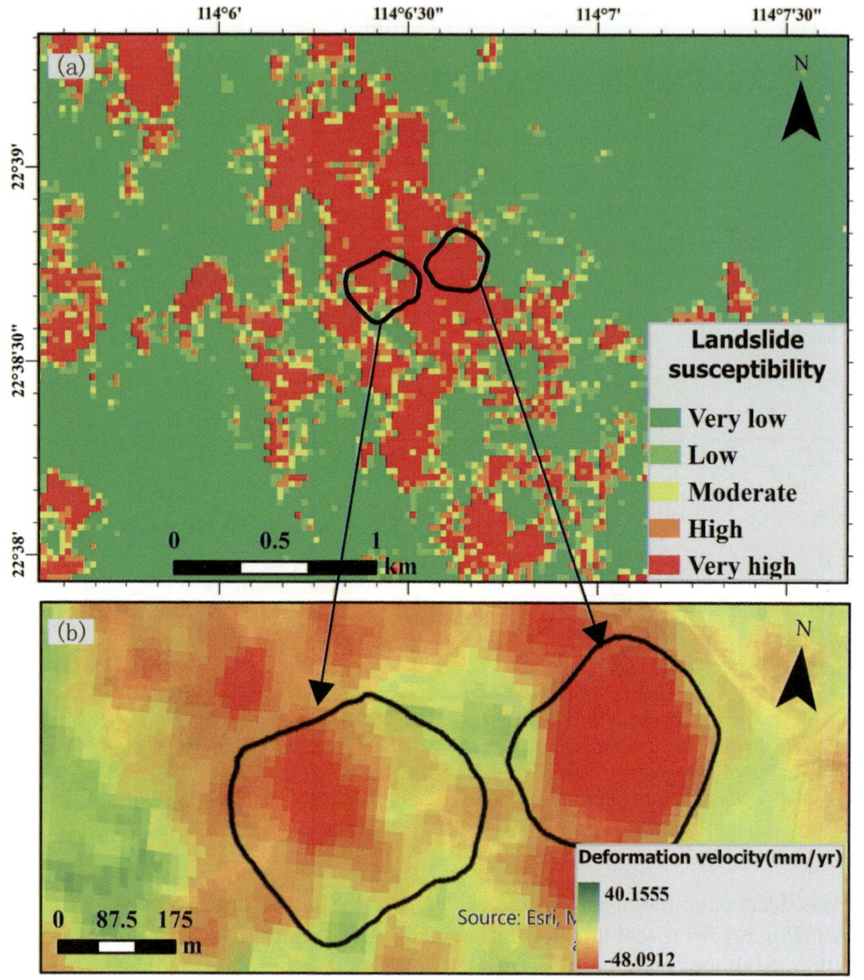

Fig. 6.27 Region 1 **a** Landslide susceptibility map, **b** InSAR based ground surface deformation map, reprinted from Ref. Gao et al. (2024), copyright 2025, with permission from IEEE

6.6 Summary

This chapter introduces a multi-model ensemble approach for landslide susceptibility mapping by integrating static and dynamic spatiotemporal data. Initially, a Stacking ensemble learning framework was developed by combining CNN, MLP, GRU, and SVR models. Model performance and the relative importance of landslide conditioning factors were thoroughly evaluated using multiple performance metrics and partial dependence plots (PDPs). Results indicate that conventional static neural network models struggle to effectively process large volumes of temporal dynamic data, leading to diminished reliability in susceptibility prediction. In contrast, the

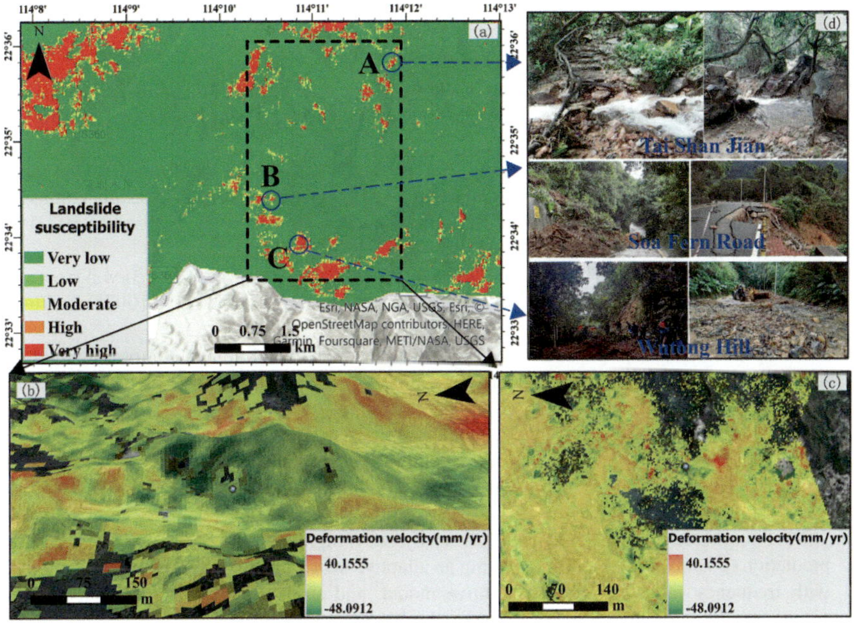

Fig. 6.28 Region 2. **a** Landslide susceptibility map, **b–c** InSAR based ground surface deformation maps, **d** landslide on September 7, 2023 at Soa Fern Road, Wutong Hill, Tai Shan Jian, Yantian District, reprinted from Ref. Gao et al. (2024), copyright 2025, with permission from IEEE

Stacking ensemble framework effectively aggregates the strengths of individual base learners, enhances overall prediction accuracy, and reduces the extent of false-positive zones. Building upon this, a TSDNN framework was proposed, leveraging a parallel architecture based on CNN for spatial feature extraction and Bi-LSTM for temporal sequence learning. This framework is capable of concurrently processing static and dynamic landslide conditioning factors, thereby enabling more efficient extraction and fusion of spatial and temporal features. Comprehensive qualitative and visualized quantitative evaluations demonstrated that, compared to the Stacking framework, the TSDNN framework offers significantly improved performance in leveraging both static and dynamic landslide-related features, thereby enhancing the reliability of susceptibility mapping. Furthermore, the predicted extremely high-risk areas derived from the TSDNN framework aligned well with actual landslide events that occurred in October 2023. An overlay analysis with InSAR based ground surface deformation rate further revealed that the majority of high-susceptibility zones coincided with areas exhibiting elevated deformation rates, underscoring the model's strong predictive consistency and practical applicability.

References

Adadi, A., & Berrada, M. (2018). Peeking inside the black-box: A survey on explainable artificial intelligence (XAI). *IEEE Access, 6*, 52138–52160. https://doi.org/10.1109/ACCESS.2018.2870052

Al-Najjar, H. A. H., Pradhan, B., Beydoun, G., Sarkar, R., Park, H.-J., & Alamri, A. (2023). A novel method using explainable artificial intelligence (XAI)-based Shapley Additive Explanations for spatial landslide prediction using Time-Series SAR dataset. *Gondwana Research, 123*, 107–124. https://doi.org/10.1016/j.gr.2022.08.004

Bouvry, P., González-Vélez, H., & Kołodziej, J. (Eds.). (2011). *Intelligent Decision Systems in Large-Scale Distributed Environments* (Vol. 362). Springer. https://doi.org/10.1007/978-3-642-21271-0

Cai, J. (2024). Satellite and ground-based InSAR time series modeling and solution method for landslide dynamic monitoring and early warning. *Acta Geodaetica Et Cartographica Sinica, 53*(1), 201.

Chen, J., Chen, X., & Yang, X. (2021). GeoDetector based research on spatial differentiation and influencing factors on china urban exhibition development. *IOP Conference Series: Earth and Environmental Science, 706*(1), Article 012014. https://doi.org/10.1088/1755-1315/706/1/012014

Chen, W., Pourghasemi, H. R., Panahi, M., Kornejady, A., Wang, J., Xie, X., & Cao, S. (2017). Spatial prediction of landslide susceptibility using an adaptive neuro-fuzzy inference system combined with frequency ratio, generalized additive model, and support vector machine techniques. *Geomorphology, 297*, 69–85. https://doi.org/10.1016/j.geomorph.2017.09.007

Ciampalini, A., Raspini, F., Lagomarsino, D., Catani, F., & Casagli, N. (2016). Landslide susceptibility map refinement using PSInSAR data. *Remote Sensing of Environment, 184*, 302–315. https://doi.org/10.1016/j.rse.2016.07.018

Cui, S., Yin, Y., Wang, D., Li, Z., & Wang, Y. (2021). A stacking-based ensemble learning method for earthquake casualty prediction. *Applied Soft Computing, 101*, Article 107038. https://doi.org/10.1016/j.asoc.2020.107038

Dai, K., Deng, J., Xu, Q., Li, Z., Shi, X., Hancock, C., Wen, N., Zhang, L., & Zhuo, G. (2022). Interpretation and sensitivity analysis of the InSAR line of sight displacements in landslide measurements. *GIScience & Remote Sensing*. https://www.tandfonline.com/doi/abs/https://doi.org/10.1080/15481603.2022.2100054

Dai, K., Chen, C., Shi, X., Wu, M., Feng, W., Xu, Q., Liang, R., Zhuo, G., & Li, Z. (2023). Dynamic landslides susceptibility evaluation in Baihetan Dam area during extensive impoundment by integrating geological model and InSAR observations. *International Journal of Applied Earth Observation and Geoinformation, 116*, Article 103157. https://doi.org/10.1016/j.jag.2022.103157

Divina, F., Gilson, A., Goméz-Vela, F., García Torres, M., & Torres, J. F. (2018). Stacking ensemble learning for short-term electricity consumption forecasting. *Energies, 11*(4), Article 4. https://doi.org/10.3390/en11040949

Fang, Z., Wang, Yi, Peng, Ling, & and Hong, H. (2021). A comparative study of heterogeneous ensemble-learning techniques for landslide susceptibility mapping. *International Journal of Geographical Information Science, 35*(2), 321–347. https://doi.org/10.1080/13658816.2020.1808897

Gao B., He Y., Zhang L., Yao S., Yang W., Chen Y., He X., Zhao Z., & Chen H. (2023). Dynamic evaluation of landslide susceptibility by CNN considering InSAR deformation: A case study of Liujiaxia reservoir. *Chinese Journal of Rock Mechanics and Engineering, 42*(2), 450–465. https://doi.org/10.13722/j.cnki.jrme.2022.0266

Gao, B., He, Y., Chen, X., Chen, H., Yang, W., & Zhang, L. (2024). A deep neural network framework for landslide susceptibility mapping by considering time-series rainfall. *IEEE Journal of Selected Topics in Applied Earth Observations and Remote Sensing, 17*, 5946–5969. https://doi.org/10.1109/JSTARS.2024.3370218

Girshick, R., Donahue, J., Darrell, T., & Malik, J. (2014). Rich feature hierarchies for accurate object detection and semantic segmentation. *IEEE Conference on Computer Vision and Pattern Recognition, 2014*, 580–587. https://doi.org/10.1109/CVPR.2014.81

He, K., H., Shangqing, W., Wen, D., & Sijing, W. (2010). Dynamic features and effects of rainfall on landslides in the Three Gorges Reservoir region, China: Using the Xintan landslide and the large Huangya landslide as the examples. *Environmental Earth Sciences, 59*(6), 1267–1274. https://doi.org/10.1007/s12665-009-0114-5

He, Y., Wang, Wenhui, Zhang, Lifeng, Chen, Youdong, Chen, Yi, Chen, Baoshan, He, Xu, & and Zhao, Z. (2023a). An identification method of potential landslide zones using InSAR data and landslide susceptibility. *Geomatics, Natural Hazards and Risk, 14*(1), 2185120. https://doi.org/10.1080/19475705.2023.2185120

Hong, H., Pourghasemi, H. R., & Pourtaghi, Z. S. (2016). Landslide susceptibility assessment in Lianhua County (China): A comparison between a random forest data mining technique and bivariate and multivariate statistical models. *Geomorphology, 259*, 105–118. https://doi.org/10.1016/j.geomorph.2016.02.012

Hong, H., Shahabi, H., Shirzadi, A., Chen, W., Chapi, K., Ahmad, B. B., Roodposhti, M. S., Yari Hesar, A., Tian, Y., & Tien Bui, D. (2019). Landslide susceptibility assessment at the Wuning area, China: A comparison between multi-criteria decision making, bivariate statistical and machine learning methods. *Natural Hazards, 96*(1), 173–212. https://doi.org/10.1007/s11069-018-3536-0

Huang, F., Yan, J., Fan, X., Yao, C., Huang, J., Chen, W., & Hong, H. (2022). Uncertainty pattern in landslide susceptibility prediction modelling: Effects of different landslide boundaries and spatial shape expressions. *Geoscience Frontiers, 13*(2), Article 101317. https://doi.org/10.1016/j.gsf.2021.101317

Ju, Y., Xu, Q., Jin, S., Li, W., Su, Y., Dong, X., & Guo, Q. (2022). Loess landslide detection using object detection algorithms in Northwest China. *Remote Sensing, 14*(5), Article 5. https://doi.org/10.3390/rs14051182

Kim, J., Coe, J. A., Lu, Z., Avdievitch, N. N., & Hults, C. P. (2022). Spaceborne InSAR mapping of landslides and subsidence in rapidly deglaciating terrain, Glacier Bay National Park and Preserve and vicinity, Alaska and British Columbia. *Remote Sensing of Environment, 281*, Article 113231. https://doi.org/10.1016/j.rse.2022.113231

Liu B., Zhang L., Ge D., Li M., Zhou X., Guo Z., Shi P., Zhang L., Jin D., Wan X., Wang Y., & Wang Y. (2024). Application of InSAR monitoring large deformation of landslides using Lutan-1 constellation. *Geomatics and Information Science of Wuhan University, 49*(10), 1753–1762. https://doi.org/10.13203/j.whugis20230478

Liu, Y., He, Y., Yang, W., & Zhang, L. (2024b). Spatio-temporal characteristics and driving mechanism of alpine peatland InSAR surface deformation—a case study of Maduo County, China. *IEEE Journal of Selected Topics in Applied Earth Observations and Remote Sensing, 17*, 4492–4514. https://doi.org/10.1109/JSTARS.2024.3359712

Luo, X., Lin, F., Zhu, S., Yu, M., Zhang, Z., Meng, L., & Peng, J. (2019). Mine landslide susceptibility assessment using IVM, ANN and SVM models considering the contribution of affecting factors. *PLoS ONE, 14*(4), Article e0215134. https://doi.org/10.1371/journal.pone.0215134

Lv, L., Chen, T., Dou, J., & Plaza, A. (2022). A hybrid ensemble-based deep-learning framework for landslide susceptibility mapping. *International Journal of Applied Earth Observation and Geoinformation, 108*, Article 102713. https://doi.org/10.1016/j.jag.2022.102713

Peng, L., Xu, S., Hou, J., & Peng, J. (2015). Quantitative risk analysis for landslides: The case of the Three Gorges area, China. *Landslides, 12*(5), 943–960. https://doi.org/10.1007/s10346-014-0518-5

Reichenbach, P., Rossi, M., Malamud, B. D., Mihir, M., & Guzzetti, F. (2018). A review of statistically-based landslide susceptibility models. *Earth-Science Reviews, 180*, 60–91. https://doi.org/10.1016/j.earscirev.2018.03.001

Rong, G., Li, K., Tong, Z., Liu, X., Zhang, J., Zhang, Y., & Li, T. (2023). Population amount risk assessment of extreme precipitation-induced landslides based on integrated machine learning model and scenario simulation. *Geoscience Frontiers, 14*(3), Article 101541. https://doi.org/10.1016/j.gsf.2023.101541

Sameen, M. I., Pradhan, B., & Lee, S. (2020). Application of convolutional neural networks featuring Bayesian optimization for landslide susceptibility assessment. *CATENA, 186*, Article 104249. https://doi.org/10.1016/j.catena.2019.104249

Tien Bui, D., Pradhan, B., Lofman, O., Revhaug, I., & Dick, O. B. (2012). Spatial prediction of landslide hazards in Hoa Binh province (Vietnam): A comparative assessment of the efficacy of evidential belief functions and fuzzy logic models. *CATENA, 96*, 28–40. https://doi.org/10.1016/j.catena.2012.04.001

Wang, J., & Xu, C. (2017). Geodetector: Principle and prospective. *Acta Geographica Sinica, 72*(1), 116–134.

Wu, T., Xie, X., Wu, H., Zeng, H., & Zhu, X. (2022). A quantitative analysis method of regional rainfall-induced landslide deformation response variation based on a time-domain correlation model. *Land, 11*(5), Article 5. https://doi.org/10.3390/land11050703

Xia, M., Shao, H., Ma, X., & de Silva, C. W. (2021). A stacked GRU-RNN-based approach for predicting renewable energy and electricity load for smart grid operation. *IEEE Transactions on Industrial Informatics, 17*(10), 7050–7059. https://doi.org/10.1109/TII.2021.3056867

Xiang, Z., Dou, J., Yunus, A. P., Zhang, L., Wang, X., Luo, W. (2023). Vegetation-landslide nexus and topographic changes post the 2004 Mw 6.6 Chuetsu earthquake. *CATENA, 223*, 106946. https://doi.org/10.1016/j.catena.2023.106946

Xing, X., Wu, C., Li, J., Li, X., Zhang, L., & He, R. (2021). Susceptibility assessment for rainfall-induced landslides using a revised logistic regression method. *Natural Hazards, 106*(1), 97–117. https://doi.org/10.1007/s11069-020-04452-4

Yang, C., Jin, Y., Deng, F., & Shi, X. (2025). Detection and monitoring of landslides along the Xuyong-Gulin Expressway using SBAS InSAR. *Remote Sensing for Natural Resources, 37*(1), 161–168.

Yi, Y., Zhang, W., Xu, X., Zhang, Z., & Wu, X. (2022). Evaluation of neural network models for landslide susceptibility assessment. *International Journal of Digital Earth, 15*(1), 934–953. https://doi.org/10.1080/17538947.2022.2062467

Yin, Y., Li, B., Wang, W., Zhan, L., Xue, Q., Gao, Y., Zhang, N., Chen, H., Liu, T., & Li, A. (2016). Mechanism of the December 2015 catastrophic landslide at the Shenzhen landfill and controlling geotechnical risks of urbanization. *Engineering, 2*(2), 230–249. https://doi.org/10.1016/J.ENG.2016.02.005

Zhang, P., Du, P., Guo, S., Zhang, W., Tang, P., Chen, J., & Zheng, H. (2022). A novel index for robust and large-scale mapping of plastic greenhouse from Sentinel-2 images. *Remote Sensing of Environment, 276*, Article 113042. https://doi.org/10.1016/j.rse.2022.113042

Zhang, Y., Meng, X. M., Dijkstra, T. A., Jordan, C. J., Chen, G., Zeng, R. Q., & Novellino, A. (2020). Forecasting the magnitude of potential landslides based on InSAR techniques. *Remote Sensing of Environment, 241*, Article 111738. https://doi.org/10.1016/j.rse.2020.111738

Zhao Z. (2024). *A Landslide Susceptibility Assessment Method with Dynamic and Static Features Combined with InSAR Deformation Information* [Lanzhou Jiaotong University]. https://doi.org/10.27205/d.cnki.gltec.2023.001728

Zhao, Z., He, Y., Yao, S., Yang, W., Wang, W., Zhang, L., & Sun, Q. (2022). A comparative study of different neural network models for landslide susceptibility mapping. *Advances in Space Research, 70*(2), 383–401. https://doi.org/10.1016/j.asr.2022.04.055

Zhou, X., Wen, H., Zhang, Y., Xu, J., & Zhang, W. (2021). Landslide susceptibility mapping using hybrid random forest with GeoDetector and RFE for factor optimization. *Geoscience Frontiers, 12*(5), Article 101211. https://doi.org/10.1016/j.gsf.2021.101211

Chapter 7
Concluding Words

This book conducts an in-depth study on intelligent landslide identification and susceptibility mapping/assessment based on InSAR and deep learning, with targeted research on three typical landslide-prone regions in China: Lanzhou and the Bailong river basin in Gansu Province, and Shenzhen in Guangdong Province.

In the field of intelligent landslide identification, the following advancements were made:

There are many issues in landslide identification based on remote sensing by traditional methods. Such as, under complex scenarios (e.g. adverse weather, vegetation, and cloud cover), relying on optical remote sensing images for deep learning-based landslide identification results in false positives, missed detections, and low accuracy, failing to meet practical needs; the issue of indistinct landslide features caused by insufficient information richness in remote sensing imagery.

We proposed multiple solutions to address this challenge: (1) multi-input channel U-Net landslide identification method integrating multi-source remote sensing data, multi-model ensemble landslide identification method incorporating spectral, (2) topographic, and texture features, and Sentinel-2A multi-band optical remote sensing images were introduced for landslide identification, constructing a channel attention mechanism-enhanced U-Net model.

Our experiments demonstrate that the proposed methods effectively leverage the rich information provided by multi-source SAR remote sensing data, with focused learning of landslide features, achieving superior performance. The approach attains high landslide identification accuracy, showing strong agreement with ground truth data. It enables more comprehensive and precise identification of landslide areas, while maintaining low false detection and omission rates.

In the field of intelligent landslide susceptibility assessment/mapping, the following advancements were made:

There are challenges in conventional landslide susceptibility assessment. Traditional landslide susceptibility assessment faces several critical limitations: data-driven machine learning methods fail to adequately characterize the spatiotemporal

Y. He, *InSAR and Deep Learning in Landslides Research: Intelligent Identification, Risk Assessment and Susceptibility Mapping*, https://doi.org/10.1007/978-981-96-9132-6_7

relationships between landslides and their predisposing environments in complex geological settings, resulting in persistent issues of sample imbalance and insufficient feature representation. Moreover, existing landslide susceptibility assessment models employ oversimplified learning paradigms with limited feature extraction capabilities, preventing them from effectively mining latent landslide-related features and adapting to heterogeneous geographical environments. Additionally, most current approaches primarily rely on static landslide factors while neglecting dynamic triggering factors, leading to low-accuracy susceptibility results and consequently unreliable landslide risk assessments.

We proposed multiple solutions to address these challenges: (1) a graph neural network landslide susceptibility mapping method constrained by geographical environment consistency, (2) integrated dynamic–static feature framework for landslide susceptibility mapping using InSAR based ground surface deformation, and (3) multi-model integrated landslide susceptibility mapping method based on static-dynamic spatiotemporal data fusion.

Experimental results demonstrate that the proposed landslide susceptibility assessment method based on geographically constrained graph neural networks significantly improves the accuracy and reliability of landslide risk evaluation. Comparative analyses confirm their superior performance over conventional methods across all evaluation metrics, effectively validating their advantages and practical value. Under the Stacking ensemble framework, our approach successfully integrates features from individual models to enhance prediction accuracy while reducing misidentified areas. Furthermore, we developed a TSDNN parallel processing framework combining CNN and Bi-LSTM architectures to jointly process static and dynamic landslide influencing factors. This innovative framework more effectively extracts and fuses the spatial and temporal dynamic characteristics of static-dynamic landslide factors. Through comprehensive qualitative evaluation using multiple metrics and quantitative visual analysis, the TSDNN framework demonstrates significantly better performance in extracting and utilizing both static and dynamic landslide features compared to the Stacking approach, markedly improving the reliability of susceptibility mapping. Notably, our method achieves more precise delineation of extremely high susceptibility zones, encompassing a greater number of historical landslide points within smaller areas, thereby producing more reliable results. The integration of these advancements establishes a robust methodological foundation for next-generation landslide early warning systems, particularly in complex terrains with active deformation processes.

GPSR Compliance

The European Union's (EU) General Product Safety Regulation (GPSR) is a set of rules that requires consumer products to be safe and our obligations to ensure this.

If you have any concerns about our products, you can contact us on ProductSafety@springernature.com

In case Publisher is established outside the EU, the EU authorized representative is:

Springer Nature Customer Service Center GmbH
Europaplatz 3
69115 Heidelberg, Germany

Batch number: 10704307

Printed by Printforce, the Netherlands